Jan Dangerfield
Stuart Haring
Series Editor: Julian Gilbey

Cambridge International
AS & A Level Mathematics:
Mechanics
Coursebook

CAMBRIDGE
UNIVERSITY PRESS

Shaftesbury Road, Cambridge CB2 8EA, United Kingdom

One Liberty Plaza, 20th Floor, New York, NY 10006, USA

477 Williamstown Road, Port Melbourne, VIC 3207, Australia

314–321, 3rd Floor, Plot 3, Splendor Forum, Jasola District Centre, New Delhi – 110025, India

103 Penang Road, #05-06/07, Visioncrest Commercial, Singapore 238467

Cambridge University Press is part of the University of Cambridge.

It furthers the University's mission by disseminating knowledge in the pursuit of education, learning and research at the highest international levels of excellence.

www.cambridge.org
Information on this title: www.cambridge.org/9781108407267

© Cambridge University Press & Assessment 2018

This publication is in copyright. Subject to statutory exception and to the provisions of relevant collective licensing agreements, no reproduction of any part may take place without the written permission of Cambridge University Press.

First published 2018

21

Printed in Great Britain by Ashford Colour Press Ltd.

A catalogue record for this publication is available from the British Library

ISBN 978-1-108-40726-7 Paperback
ISBN 978-1-108-56294-2 Paperback + Cambridge Online Mathematics, 2 years
ISBN 978-1-108-46223-5 Cambridge Online Mathematics, 2 years

Cambridge University Press has no responsibility for the persistence or accuracy of URLs for external or third-party internet websites referred to in this publication, and does not guarantee that any content on such websites is, or will remain, accurate or appropriate. Information regarding prices, travel timetables, and other factual information given in this work is correct at the time of first printing but Cambridge University Press does not guarantee the accuracy of such information thereafter.

®*IGCSE is a registered trademark*

Past exam paper questions throughout are reproduced by permission of Cambridge Assessment International Education. Cambridge Assessment International Education bears no responsibility for the example answers to questions taken from its past question papers which are contained in this publication. The questions, example answers, marks awarded and/or comments that appear in this book were written by the author(s). In examination, the way marks would be awarded to answers like these may be different.

NOTICE TO TEACHERS IN THE UK
It is illegal to reproduce any part of this work in material form (including photocopying and electronic storage) except under the following circumstances:
(i) where you are abiding by a licence granted to your school or institution by the Copyright Licensing Agency;
(ii) where no such licence exists, or where you wish to exceed the terms of a licence, and you have gained the written permission of Cambridge University Press;
(iii) where you are allowed to reproduce without permission under the provisions of Chapter 3 of the Copyright, Designs and Patents Act 1988, which covers, for example, the reproduction of short passages within certain types of educational anthology and reproduction for the purposes of setting examination questions.

Contents

Series introduction	vi
How to use this book	viii
Acknowledgements	x
1 Velocity and acceleration	**1**
1.1 Displacement and velocity	2
1.2 Acceleration	9
1.3 Equations of constant acceleration	11
1.4 Displacement–time graphs and multi-stage problems	15
1.5 Velocity–time graphs and multi-stage problems	22
1.6 Graphs with discontinuities	27
End-of-chapter review exercise 1	31
2 Force and motion in one dimension	**35**
2.1 Newton's first law and relation between force and acceleration	36
2.2 Combinations of forces	39
2.3 Weight and motion due to gravity	42
2.4 Normal contact force and motion in a vertical line	47
End-of-chapter review exercise 2	51
3 Forces in two dimensions	**53**
3.1 Resolving forces in horizontal and vertical directions in equilibrium problems	54
3.2 Resolving forces at other angles in equilibrium problems	59
Ⓔ 3.3 The triangle of forces and Lami's theorem for three-force equilibrium problems	63
3.4 Non-equilibrium problems for objects on slopes and known directions of acceleration	67
3.5 Non-equilibrium problems and finding resultant forces and directions of acceleration	72
End-of-chapter review exercise 3	78
Cross-topic review exercise 1	**81**

4 Friction — 83
- 4.1 Friction as part of the contact force — 84
- 4.2 Limit of friction — 91
- 4.3 Change of direction of friction in different stages of motion — 96
- **E** 4.4 Angle of friction — 101
- End-of-chapter review exercise 4 — 106

5 Connected particles — 109
- 5.1 Newton's third law — 110
- 5.2 Objects connected by rods — 111
- 5.3 Objects connected by strings — 116
- 5.4 Objects in moving lifts (elevators) — 122
- End-of-chapter review exercise 5 — 127

6 General motion in a straight line — 130
- 6.1 Velocity as the derivative of displacement with respect to time — 132
- 6.2 Acceleration as the derivative of velocity with respect to time — 136
- 6.3 Displacement as the integral of velocity with respect to time — 141
- 6.4 Velocity as the integral of acceleration with respect to time — 150
- End-of-chapter review exercise 6 — 155

Cross-topic review exercise 2 — 157

7 Momentum — 159
- 7.1 Momentum — 161
- 7.2 Collisions and conservation of momentum — 163
- End-of-chapter review exercise 7 — 170

8 Work and energy — 172
- 8.1 Work done by a force — 174
- 8.2 Kinetic energy — 180
- 8.3 Gravitational potential energy — 183
- End-of-chapter review exercise 8 — 186

9 The work–energy principle and power	**188**
9.1 The work–energy principle	189
9.2 Conservation of energy in a system of conservative forces	196
9.3 Conservation of energy in a system with non-conservative forces	199
9.4 Power	204
End-of-chapter review exercise 9	209
Cross-topic review exercise 3	211
Practice exam-style paper	213
Answers	215
Glossary	235
Index	237

Series introduction

Cambridge International AS & A Level Mathematics can be a life-changing course. On the one hand, it is a facilitating subject: there are many university courses that either require an A Level or equivalent qualification in mathematics or prefer applicants who have it. On the other hand, it will help you to learn to think more precisely and logically, while also encouraging creativity. Doing mathematics can be like doing art: just as an artist needs to master her tools (use of the paintbrush, for example) and understand theoretical ideas (perspective, colour wheels and so on), so does a mathematician (using tools such as algebra and calculus, which you will learn about in this course). But this is only the technical side: the joy in art comes through creativity, when the artist uses her tools to express ideas in novel ways. Mathematics is very similar: the tools are needed, but the deep joy in the subject comes through solving problems.

You might wonder what a mathematical 'problem' is. This is a very good question, and many people have offered different answers. You might like to write down your own thoughts on this question, and reflect on how they change as you progress through this course. One possible idea is that a mathematical problem is a mathematical question that you do not immediately know how to answer. (If you do know how to answer it immediately, then we might call it an 'exercise' instead.) Such a problem will take time to answer: you may have to try different approaches, using different tools or ideas, on your own or with others, until you finally discover a way into it. This may take minutes, hours, days or weeks to achieve, and your sense of achievement may well grow with the effort it has taken.

In addition to the mathematical tools that you will learn in this course, the problem-solving skills that you will develop will also help you throughout life, whatever you end up doing. It is very common to be faced with problems, be it in science, engineering, mathematics, accountancy, law or beyond, and having the confidence to systematically work your way through them will be very useful.

This series of Cambridge International AS & A Level Mathematics coursebooks, written for the Cambridge Assessment International Education syllabus for examination from 2020, will support you both to learn the mathematics required for these examinations and to develop your mathematical problem-solving skills. The new examinations may well include more unfamiliar questions than in the past, and having these skills will allow you to approach such questions with curiosity and confidence.

In addition to problem solving, there are two other key concepts that Cambridge Assessment International Education have introduced in this syllabus: namely communication and mathematical modelling. These appear in various forms throughout the coursebooks.

Communication in speech, writing and drawing lies at the heart of what it is to be human, and this is no less true in mathematics. While there is a temptation to think of mathematics as only existing in a dry, written form in textbooks, nothing could be further from the truth: mathematical communication comes in many forms, and discussing mathematical ideas with colleagues is a major part of every mathematician's working life. As you study this course, you will work on many problems. Exploring them or struggling with them together with a classmate will help you both to develop your understanding and thinking, as well as improving your (mathematical) communication skills. And being able to convince someone that your reasoning is correct, initially verbally and then in writing, forms the heart of the mathematical skill of 'proof'.

Mathematical modelling is where mathematics meets the 'real world'. There are many situations where people need to make predictions or to understand what is happening in the world, and mathematics frequently provides tools to assist with this. Mathematicians will look at the real world situation and attempt to capture the key aspects of it in the form of equations, thereby building a model of reality. They will use this model to make predictions, and where possible test these against reality. If necessary, they will then attempt to improve the model in order to make better predictions. Examples include weather prediction and climate change modelling, forensic science (to understand what happened at an accident or crime scene), modelling population change in the human, animal and plant kingdoms, modelling aircraft and ship behaviour, modelling financial markets and many others. In this course, we will be developing tools which are vital for modelling many of these situations.

To support you in your learning, these coursebooks have a variety of new features, for example:

- Explore activities: These activities are designed to offer problems for classroom use. They require thought and deliberation: some introduce a new idea, others will extend your thinking, while others can support consolidation. The activities are often best approached by working in small groups and then sharing your ideas with each other and the class, as they are not generally routine in nature. This is one of the ways in which you can develop problem-solving skills and confidence in handling unfamiliar questions.
- Questions labelled as (P), (M) or (PS): These are questions with a particular emphasis on 'Proof', 'Modelling' or 'Problem solving'. They are designed to support you in preparing for the new style of examination. They may or may not be harder than other questions in the exercise.
- The language of the explanatory sections makes much more use of the words 'we', 'us' and 'our' than in previous coursebooks. This language invites and encourages you to be an active participant rather than an observer, simply following instructions ('you do this, then you do that'). It is also the way that professional mathematicians usually write about mathematics. The new examinations may well present you with unfamiliar questions, and if you are used to being active in your mathematics, you will stand a better chance of being able to successfully handle such challenges.

At various points in the books, there are also web links to relevant Underground Mathematics resources, which can be found on the free **undergroundmathematics.org** website. Underground Mathematics has the aim of producing engaging, rich materials for all students of Cambridge International AS & A Level Mathematics and similar qualifications. These high-quality resources have the potential to simultaneously develop your mathematical thinking skills and your fluency in techniques, so we do encourage you to make good use of them.

We wish you every success as you embark on this course.

Julian Gilbey
London, 2018

Past exam paper questions throughout are reproduced by permission of Cambridge Assessment International Education. Cambridge Assessment International Education bears no responsibility for the example answers to questions taken from its past question papers which are contained in this publication.

The questions, example answers, marks awarded and/or comments that appear in this book were written by the author(s). In examination, the way marks would be awarded to answers like these may be different.

How to use this book

Throughout this book you will notice particular features that are designed to help your learning. This section provides a brief overview of these features.

In this chapter you will learn how to:
- use Newton's third law for objects that are in contact
- calculate the motion or equilibrium of objects connected by rods
- calculate the motion or equilibrium of objects connected by strings
- calculate the motion or equilibrium of objects that are moving in elevators.

Learning objectives indicate the important concepts within each chapter and help you to navigate through the coursebook.

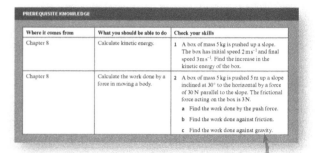

Prerequisite knowledge exercises identify prior learning that you need to have covered before starting the chapter. Try the questions to identify any areas that you need to review before continuing with the chapter.

KEY POINT 5.1

In a connected system, you can apply Newton's second law to the entire system or to the individual components of the system.

Key point boxes contain a summary of the most important methods, facts and formulae.

instantaneous velocity

Key terms are important terms in the topic that you are learning. They are highlighted in orange bold. The **glossary** contains clear definitions of these key terms.

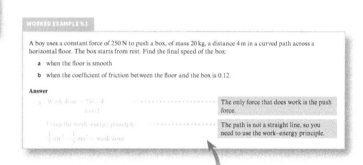

Worked examples provide step-by-step approaches to answering questions. The left side shows a fully worked solution, while the right side contains a commentary explaining each step in the working.

EXPLORE 6.3

A particle moves in a straight line so that its acceleration at time t seconds is $a\,\mathrm{m\,s^{-2}}$, where a is constant. The initial velocity of the particle is $u\,\mathrm{m\,s^{-1}}$.

Use calculus to find the velocity of the particle as a function of t.

Explore boxes contain enrichment activities for extension work. These activities promote group work and peer-to-peer discussion, and are intended to deepen your understanding of a concept. (Answers to the Explore questions are provided in the Teacher's Resource.)

TIP

It is normally easier to put all the information in a question into this equation rather than working out net force separately.

Tip boxes contain helpful guidance about calculating or checking your answers.

How to use this book

REWIND

Recall, from Chapter 6, that when we differentiate displacement with respect to time we get velocity, and when we differentiate distance with respect to time we get speed.

FAST FORWARD

Newton's third law relates forces between objects and their effects on each other. You will learn about this in Chapter 5.

Extension material goes beyond the syllabus. It is highlighted by a red line to the left of the text.

Web link boxes contain links to useful resources on the internet.

WEB LINK

You may want to have a go at the *Make it stop* resource at the *Vector Geometry* station on the Underground Mathematics website.

Rewind and **Fast forward** boxes direct you to related learning. **Rewind** boxes refer to earlier learning, in case you need to revise a topic. **Fast forward** boxes refer to topics that you will cover at a later stage, in case you would like to extend your study.

MODELLING ASSUMPTIONS

In practice, the objects may not instantaneously change velocity. In the example of a tennis ball being hit by a racket, the strings stretch very slightly and spring back into shape. It is during this time that the ball changes velocity. In the case of a tennis ball striking a solid wall or a solid object striking the ball, the ball may compress slightly during contact before springing back into shape. In these cases, the time required to change is so small that you can ignore it. By modelling the objects as particles, you can assume the objects do not lose shape and the time in contact is sufficiently small to be negligible.

The real world doesn't often provide you with a straightforward question to answer and instead requires you to make assumptions in order to model situations. The **Modelling assumptions** box describes the important assumptions that have been made in the topic.

Throughout each chapter there are multiple exercises containing practice questions. The questions are coded:

PS These questions focus on problem-solving.

P These questions focus on proofs.

M These questions focus on modelling.

✏ You should not use a calculator for these questions.

▣ These questions are taken from past examination papers.

The **End-of-chapter review** contains exam-style questions covering all topics in the chapter. You can use this to check your understanding of the topics you have covered. The number of marks gives an indication of how long you should be spending on the question. You should spend more time on questions with higher mark allocations; questions with only one or two marks should not need you to spend time doing complicated calculations or writing long explanations.

DID YOU KNOW?

Galileo Galilei (1564–1642) was the first to demonstrate that the mass does not affect the acceleration in free fall. It was thought he did this by dropping balls of the same material but of different masses from the Leaning Tower of Pisa to show they land at the same time. However, no account of this was made by Galileo and it is generally considered to have been a thought experiment. Actual experiments on inclined planes did verify Galileo's theory.

Did you know? boxes contain interesting facts showing how Mathematics relates to the wider world.

Checklist of learning and understanding

- A force is something that influences the motion of an object. Its size is measured in newtons (N).
- Force is related to acceleration by the equation: net force = mass × acceleration.
- Objects acted on only by the force of gravity have an acceleration of g ms^{-2}.
- The weight of an object is the force on it due to gravity and has magnitude $W = mg$.
- The reaction force or normal contact force is the force on an object due to being in contact with another object or surface. It acts perpendicular to the surface and is usually denoted by R (or sometimes N).

At the end of each chapter there is a **Checklist of learning and understanding**. The checklist contains a summary of the concepts that were covered in the chapter. You can use this to quickly check that you have covered the main topics.

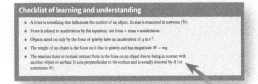

Cross-topic review exercises appear after several chapters, and cover topics from across the preceding chapters.

Acknowledgements

The authors and publishers acknowledge the following sources of copyright material and are grateful for the permissions granted. While every effort has been made, it has not always been possible to identify the sources of all the material used, or to trace all copyright holders. If any omissions are brought to our notice, we will be happy to include the appropriate acknowledgements on reprinting.

Past paper exam questions throughout are reproduced by permission of Cambridge Assessment International Education.

Thanks to the following for permission to reproduce image:

Cover image malerapaso/Getty Images

Inside (*in order of appearance*) claudio.arnese/Getty Images; GJON MILI/Getty Images; Oliver Furrer/Getty Images; VCG Wilson/Corbis via Getty Images; Kean Collection /Getty Images; Julen Garces Carro/EyeEm/Getty Images; powerofforever/Getty Images; Seb Oliver/Getty Images; Buena Vista Images/Getty Images; Education Images/UIG/Getty Images; Mike Chew/Getty Images; Stock Montage / Getty Images; PAUL ELLIS / AFP/Getty Images; Blend Images - FreamPictures/Shannon Faulk/Getty Images; CTRPhotos/Getty Images

Chapter 1
Velocity and acceleration

In this chapter you will learn how to:

- work with scalar and vector quantities for distance and speed
- use equations of constant acceleration
- sketch and read displacement–time graphs and velocity–time graphs
- solve problems with multiple stages of motion.

PREREQUISITE KNOWLEDGE

Where it comes from	What you should be able to do	Check your skills
IGCSE® / O Level Mathematics	Solve quadratics by factorising or using the quadratic formula.	1 Solve the following equations. a $x^2 - 2x - 15 = 0$ b $2x^2 + x - 3 = 0$ c $3x^2 - 5x - 7 = 0$
IGCSE / O Level Mathematics	Solve linear simultaneous equations.	2 Solve the following pairs of simultaneous equations. a $2x + 3y = 8$ and $5x - 2y = 1$ b $3x + 2y = 9$ and $y = 4x - 1$

What is Mechanics about?

How far should the driver of a car stay behind another car to be able to stop safely in an emergency? How long should the fuse on a firework be so the firework goes off at the highest point? How quickly should you roll a ball so it stops as near as possible to a target? How strong does a building have to be to survive a hurricane? Mechanics is the study of questions such as these. By modelling situations mathematically and making suitable assumptions you can find answers to these questions.

In this chapter, you will study the motion of objects and learn how to work out where an object is and how it is moving at different times. This area of Mechanics is known as 'dynamics'. Solving problems with objects that do not move is called 'statics'; you will study this later in the course.

1.1 Displacement and velocity

An old English nursery rhyme goes like this:

> The Grand Old Duke of York,
>
> He had ten thousand men,
>
> He marched them up to the top of the hill,
>
> And he marched them down again.

His men had clearly marched some distance, but they ended up exactly where they started, so you cannot work out how far they travelled simply by measuring how far their finishing point is from their starting point.

You can use two different measures when thinking about how far something has travelled. These are **distance** and **displacement**.

Distance is a **scalar** quantity and is used to measure the total length of path travelled. In the rhyme, if the distance covered up the hill were 100 m, the total distance in marching up the hill and then down again would be $100\,\text{m} + 100\,\text{m} = 200\,\text{m}$.

Chapter 1: Velocity and acceleration

Displacement is a **vector** quantity and gives the location of an object relative to a fixed reference point or **origin**. In this course, you will be considering dynamics problems in only one dimension. To define the displacement you need to define one direction as positive. In the rhyme, if you take the origin to be the bottom of the hill and the positive direction to be up the hill, then the displacement at the end is 0 m, since the men are in the same location as they started. You can also reach this answer through a calculation. If you assume that they are marching in a straight line, then marching up the hill is an increase in displacement and marching down the hill is a decrease in displacement, so the total displacement is $(+100 \text{ m}) + (-100 \text{ m}) = 0 \text{ m}$.

Since you will be working in only one dimension, you will often refer to the displacement as just a number, with positive meaning a displacement in one direction from the origin and negative meaning a displacement in the other direction. Sometimes the direction and origin will be stated in the problem. In other cases, you will need to choose these yourself. In many cases the origin will simply be the starting position of an object and the positive direction will be the direction the object is moving initially.

> **TIP**
>
> A scalar quantity, such as distance, has only a magnitude. A vector quantity, such as displacement, has magnitude and direction. When you are asked for a vector quantity such as displacement or velocity, make sure you state the direction as well as the magnitude.

> **KEY POINT 1.1**
>
> Displacement is a measure of location from a fixed origin or starting point. It is a vector and so has both magnitude and direction. If you take displacement in a given direction to be positive, then displacement in the opposite direction is negative.

We also have two ways to measure how quickly an object is moving: **speed** and **velocity**. Speed is a scalar quantity, so has only a magnitude. Velocity is a vector quantity, so has both magnitude and direction.

For an object moving at constant speed, if you know the distance travelled in a given time you can work out the speed of the object.

> **WEB LINK**
>
> Try the *Discussing distance* resource at the *Introducing calculus* station on the Underground Mathematics website (www.underground mathematics.org).

> **KEY POINT 1.2**
>
> For an object moving at constant speed:
>
> $$\text{speed} = \frac{\text{distance covered}}{\text{time taken}}$$

This is valid only for objects moving at constant speed. For objects moving at non-constant speed you can consider the average speed.

> **KEY POINT 1.3**
>
> $$\text{average speed} = \frac{\text{total distance covered}}{\text{total time taken}}$$

Velocity measures how quickly the displacement of an object changes. You can write an equation similar to the one for speed.

Cambridge International AS & A Level Mathematics: Mechanics

KEY POINT 1.4

For an object moving at constant velocity:
$$\text{velocity} = \frac{\text{change in displacement}}{\text{time taken}}$$

Let's see what this means in practice.

Suppose a man is doing a fitness test. In each stage of the test he runs backwards and forwards along the length of a small football pitch. He starts at the centre spot, runs to one end of the pitch, changes direction and runs to the other end, changes direction and runs back to the centre spot, as shown in the diagrams. He runs at $4\,\text{m}\,\text{s}^{-1}$ and the pitch is 40 m long.

To define displacement and velocity you will need to define the origin and the direction you will call positive. Let's call the centre spot the origin and to the right as positive.

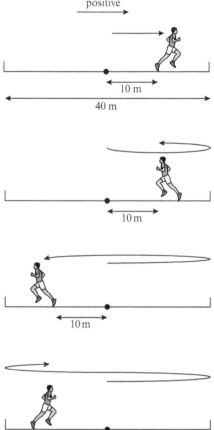

In the first diagram, he has travelled a distance of 10 m. Because he is 10 m in the positive direction, his displacement is 10 m. His speed is $4\,\text{m}\,\text{s}^{-1}$. Because he is moving in the positive direction, his velocity is also $4\,\text{m}\,\text{s}^{-1}$.

In the second diagram, he has travelled a total distance of 30 m, but he is only 10 m from the centre spot, so his displacement is 10 m. His speed is still $4\,\text{m}\,\text{s}^{-1}$ but he is moving in the negative direction so his velocity is $-4\,\text{m}\,\text{s}^{-1}$.

In the third diagram, he has travelled a total distance of 50 m, but he is now 10 m from the centre spot in the negative direction, so his displacement is -10 m. His speed is still $4\,\text{m}\,\text{s}^{-1}$ and he is still moving in the negative direction so his velocity is still $-4\,\text{m}\,\text{s}^{-1}$.

In the fourth diagram, he has travelled a total distance of 70 m, but his displacement is still -10 m. His speed is still $4\,\text{m}\,\text{s}^{-1}$ and he is moving in the positive direction again so his velocity is also $4\,\text{m}\,\text{s}^{-1}$.

The magnitude of the velocity of an object is its speed. Speed can never be negative. For example, an object moving with a velocity of $+10\,\text{m}\,\text{s}^{-1}$ and an object moving with a velocity of $-10\,\text{m}\,\text{s}^{-1}$ both have a speed of $10\,\text{m}\,\text{s}^{-1}$.

As with speed, for objects moving at non-constant velocity you can consider the average velocity.

KEY POINT 1.5

$$\text{average velocity} = \frac{\text{net change in displacement}}{\text{total time taken}}$$

> **TIP**
>
> We use vertical lines to indicate magnitude of a vector.
>
> So, $\text{speed} = |\text{velocity}|$

In the previous example, the man's average speed is $4\,\mathrm{m\,s^{-1}}$ but his average velocity is $0\,\mathrm{m\,s^{-1}}$.

We can rearrange the equation for velocity to deduce that for an object moving at constant velocity v for time t, the change in displacement s (in the same direction as the velocity) is given by:

$$s = vt$$

The standard units used for distance and displacement are metres (m) and for time are seconds (s). Therefore, the units for speed and velocity are metres per second (usually written in mathematics and science as $\mathrm{m\,s^{-1}}$, although you may also come across the notation m/s). These units are those specified by the *Système Internationale* (SI), which defines the system of units used by scientists all over the world. Other commonly used units for speed include kilometres per hour (km/h) and miles per hour (mph).

 WEB LINK

Try the *Speed vs velocity* resource at the *Introducing calculus* station on the Underground Mathematics website.

WORKED EXAMPLE 1.1

A car travels 9 km in 15 minutes at constant speed. Find its speed in $\mathrm{m\,s^{-1}}$.

Answer

$9\,\mathrm{km} = 9000\,\mathrm{m}$ and   Convert to units required for the answer, which are SI units.
$15\,\mathrm{minutes} = 900\,\mathrm{s}$

$s = vt$ Substitute into the equation for displacement and solve.
so $9000 = 900v$

$v = 10\,\mathrm{m\,s^{-1}}$

TIP

You usually only include units in the final answer to a problem and not in all the earlier steps. This is because it is easy to confuse units and variables. For example, s for displacement can be easily mixed up with s for seconds. It is important to work in SI units throughout, so that the units are consistent.

WORKED EXAMPLE 1.2

A cyclist travels at $5\,\mathrm{m\,s^{-1}}$ for $30\,\mathrm{s}$ then turns back, travelling at $3\,\mathrm{m\,s^{-1}}$ for $10\,\mathrm{s}$. Find her displacement in the original direction of motion from her starting position.

Answer

$s = vt$
So $s_1 = 5 \times 30$ Separate the two stages of the journey.
$= 150$

and $s_2 = -3 \times 10$ Remember travelling back means a negative velocity and a negative displacement.
$= -30$

So the total displacement is
$s = 150 + (-30)$
$= 120\,\mathrm{m}$

Cambridge International AS & A Level Mathematics: Mechanics

WORKED EXAMPLE 1.3

A cyclist spends some of his journey going downhill at $15\,\text{m s}^{-1}$ and the rest of the time going uphill at $5\,\text{m s}^{-1}$. In 1 minute he travels 540 m. Find how long he spent going downhill.

Answer

Let t be the amount of time spent going downhill. .... Define the variable.

Then $60 - t$ is the amount of time spent going uphill. .... Write an expression for the time spent travelling uphill.

Total distance $= 15t + 5(60 - t) = 540$ .... Set up an equation for the total distance.

$15t + 300 - 5t = 540$

$10t = 240$

$t = 24\,\text{s}$

EXPLORE 1.1

Two students are trying to solve this puzzle.

A cyclist cycles from home uphill to the shop at $5\,\text{m s}^{-1}$. He then cycles home and wants to average $10\,\text{m s}^{-1}$ for the total journey. How fast must he cycle on the way home?

The students' solutions are shown here. Decide whose logic is correct and try to explain what is wrong with the other's answer.

Student A	Student B
Call the speed on the return journey v. The average of $5\,\text{m s}^{-1}$ and v is $10\,\text{m s}^{-1}$, so v must be $15\,\text{m s}^{-1}$.	Cycling at $5\,\text{m s}^{-1}$ will take twice as long as it would if he were going at $10\,\text{m s}^{-1}$. That means he has used up the time required to go there and back in the first part of the journey, so it is impossible to average $10\,\text{m s}^{-1}$ for the total journey.

 WEB LINK

You may want to have a go at the *Average speed* resource at the *Introducing calculus* station on the Underground Mathematics website.

MODELLING ASSUMPTIONS

Throughout this course, there will be questions about how realistic your answers are. To simplify problems you will make reasonable assumptions about the scenario to allow you to solve them to a satisfactory degree of accuracy. To improve the agreement of your model with what happens in the real world, you would need to refine your model, taking into account factors that you had initially ignored.

In some of the questions so far, you might ask if it is reasonable to assume constant speed. In real life, speed would always change slightly, but it could be close enough to constant that it is a reasonable assumption.

With real objects, such as bicycles or cars, there is the question of which part of the object you are referring to. You can be consistent and say it is the front of a vehicle, but when it is a person the front changes from the left leg to the right leg. You may choose to consider the position of the torso as the position of the person. In all the examples in this coursebook, you will consider the object to be a particle, which is very small, so you do not need to worry about these details. You will assume any resulting errors in the calculations will be sufficiently small to ignore. This could cause a problem when you consider the gap between objects, because you may not have allowed for the length of the object itself, but in our simple models you will ignore this issue too.

DID YOU KNOW?

Once they have reached top speed, swimmers tend to move at a fairly constant speed at all points during the stroke. However, the race ends when the swimmer touches the end of the pool, so it is important to time the last two or three strokes to finish with arms extended. If the stroke finishes early the swimmer might not do another stroke and instead keep their arms extended, but this means the swimmer slows down. In a close race, another swimmer may overtake if that swimmer times their strokes better. This happened to Michael Phelps when he lost to Chad Le Clos in the final of the Men's 200 m Butterfly in the 2012 London Olympics.

EXERCISE 1A

1. A cyclist covers 120 m in 15 s at constant speed. Find her speed.

2. A sprinter runs at constant speed of $9 \, \text{m s}^{-1}$ for 7 s. Find the distance covered.

3. **a** A cheetah spots a grazing gazelle 150 m away and runs at a constant $25 \, \text{m s}^{-1}$ to catch it. Find how long the cheetah takes to catch the gazelle.

   b What assumptions have been made to answer the question?

4. The speed of light is $3.00 \times 10^8 \, \text{m s}^{-1}$ to 3 significant figures. The average distance between the Earth and the Sun is 150 million km to 3 significant figures. Find how long it takes for light from the Sun to reach the Earth on average. Give the answer in minutes and seconds.

5. The land speed record was set in 1997 at $1223.657 \, \text{km h}^{-1}$. Find how long in seconds it took to cover 1 km when the record was set.

Cambridge International AS & A Level Mathematics: Mechanics

6 A runner runs at $5\,\text{m s}^{-1}$ for $7\,\text{s}$ before increasing the pace to $7\,\text{m s}^{-1}$ for the next $13\,\text{s}$.

   a Find her average speed.

   b What assumptions have been made to answer the question?

7 A remote control car travels forwards at $6\,\text{m s}^{-1}$ in Drive and backwards at $3\,\text{m s}^{-1}$ in Reverse. The car travels for $10\,\text{s}$ in Drive before travelling for $5\,\text{s}$ in Reverse.

   a Find its displacement from its starting point.

   b Find its average velocity in the direction in which it started driving forwards.

   c Find its average speed.

8 A speed skater averages $11\,\text{m s}^{-1}$ over the first $5\,\text{s}$ of a race. Find the average speed required over the next $10\,\text{s}$ to average $12\,\text{m s}^{-1}$ overall.

9 The speed of sound in wood is $3300\,\text{m s}^{-1}$ and the speed of sound in air is $330\,\text{m s}^{-1}$. A hammer hits one end of a $33\,\text{m}$ long plank of wood. Find the difference in time between the sound waves being detected at the other end of the plank and the sound being heard through the air.

10 An exercise routine involves a mixture of jogging at $4\,\text{m s}^{-1}$ and sprinting at $7\,\text{m s}^{-1}$. An athlete covers $1\,\text{km}$ in 3 minutes and 10 seconds. Find how long she spent sprinting.

11 Two cars are racing over the same distance. They start at the same time, but one finishes $8\,\text{s}$ before the other. The faster one averaged $45\,\text{m s}^{-1}$ and the slower one averaged $44\,\text{m s}^{-1}$. Find the length of the race.

12 Two air hockey pucks are $2\,\text{m}$ apart. One is struck and moves directly towards the other at $1.3\,\text{m s}^{-1}$. The other is struck $0.2\,\text{s}$ later and moves directly towards the first at $1.7\,\text{m s}^{-1}$. Find how far the first puck has moved when the collision occurs and how long it has been moving for.

13 A motion from point A to point C is split into two parts. The motion from A to B has displacement s_1 and takes time t_1. The motion from B to C has displacement s_2 and takes time t_2.

   a Prove that if $t_1 = t_2$, the average speed from A to C is the same as the average of the speeds from A to B and from B to C.

   b Prove that if $s_1 = s_2$, the average speed from A to C is the same as the average of the speeds from A to B and from B to C if, and only if, $t_1 = t_2$.

14 The distance from point A to point B is s. In the motion from A to B and back, the speed for the first part of the motion is v_1 and the speed for the return part of the motion is v_2. The average speed for the entire motion is v.

   a Prove that $v = \dfrac{2v_1 v_2}{v_1 + v_2}$.

   b Deduce that it is impossible to average twice the speed of the first part of the motion; that is, it is impossible to have $v = 2v_1$.

Chapter 1: Velocity and acceleration

1.2 Acceleration

Velocity is not the only measure of the motion of an object. It is useful to know if, and how, the velocity is changing. We use **acceleration** to measure how quickly velocity is changing.

> **KEY POINT 1.6**
>
> For an object moving at constant acceleration,
>
> $$\text{acceleration} = \frac{\text{change in velocity}}{\text{time taken}}$$
>
> If an object has constant acceleration a, initial velocity u and it reaches final velocity v in time t, then
>
> $$a = \frac{v - u}{t}$$
>
> where u, v and a are all measured in the same direction.

> **TIP**
>
> The units of acceleration are m s^{-2}.

An increase in velocity is a positive acceleration, as shown in the diagram on the left.

A decrease in velocity is a negative acceleration, as shown in the diagram on the right. This is often termed a deceleration.

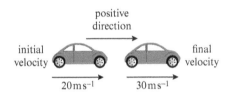
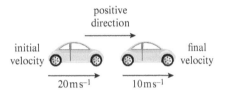

> **EXPLORE 1.2**
>
> If the initial velocity is negative, what effect would a positive acceleration have on the car? Would it be moving more quickly or less quickly?
>
> What effect would a negative acceleration have on the car in this situation? Would it be moving more quickly or less quickly?

When the acceleration is constant, the average velocity is simply the average of the initial and final velocities, which is given by the formula $\frac{1}{2}(u + v)$. This can be used to find displacements using the equation for average velocity from Key Point 1.5.

> **KEY POINT 1.7**
>
> If an object has constant acceleration a, initial velocity u and it reaches final velocity v in time t, then the displacement s is given by
>
> $$s = \frac{1}{2}(u + v)t$$

WORKED EXAMPLE 1.4

A parachutist falls from rest to $49\,\text{m s}^{-1}$ over $5\,\text{s}$. Find her acceleration.

Answer

$a = \dfrac{v - u}{t}$

$= \dfrac{49 - 0}{5}$

$= 9.8\,\text{m s}^{-2}$

Make sure you use the correct units, which are m s^{-2}.

TIP

'Rest' means not moving, so velocity is zero.

WORKED EXAMPLE 1.5

A tractor accelerates from $5\,\text{m s}^{-1}$ to $9\,\text{m s}^{-1}$ at $0.5\,\text{m s}^{-2}$. Find the distance covered by the tractor over this time.

Answer

$a = \dfrac{v - u}{t}$

So $0.5 = \dfrac{9 - 5}{t}$

$0.5t = 4$

$t = 8\,\text{s}$

Substitute into $a = \dfrac{v - u}{t}$ first to find t.

$s = \dfrac{1}{2}(u + v)t$

$= \dfrac{1}{2}(5 + 9) \times 8$

$= 56\,\text{m}$

Substitute into $s = \dfrac{1}{2}(u + v)t$ to find s.

EXERCISE 1B

1 A car accelerates from $4\,\text{m s}^{-1}$ to $10\,\text{m s}^{-1}$ in $3\,\text{s}$ at constant acceleration. Find its acceleration.

2 A car accelerates from rest to $10\,\text{m s}^{-1}$ in $4\,\text{s}$ at constant acceleration. Find its acceleration.

3 A car accelerates from $3\,\text{m s}^{-1}$ at an acceleration of $6\,\text{m s}^{-2}$. Find the time taken to reach $12\,\text{m s}^{-1}$.

4 An aeroplane accelerates at a constant rate of $3\,\text{m s}^{-2}$ for $5\,\text{s}$ from an initial velocity of $4\,\text{m s}^{-1}$. Find its final velocity.

5 A speedboat accelerates at a constant rate of $1.5\,\text{m s}^{-2}$ for $4\,\text{s}$, reaching a final velocity of $9\,\text{m s}^{-1}$. Find its initial velocity.

6 A car decelerates at a constant rate of $2\,\text{m s}^{-2}$ for $3\,\text{s}$, finishing at a velocity of $8\,\text{m s}^{-1}$. Find its initial velocity.

Chapter 1: Velocity and acceleration

7 A car accelerates from an initial velocity of $4\,\text{m}\,\text{s}^{-1}$ to a final velocity of $8\,\text{m}\,\text{s}^{-1}$ at a constant rate of $0.5\,\text{m}\,\text{s}^{-2}$. Find the car's displacement in that time.

8 A sprinter covers $60\,\text{m}$ in $10\,\text{s}$ accelerating from a jog. Her final velocity is $9\,\text{m}\,\text{s}^{-1}$.

  a Calculate her acceleration.

  b What assumptions have been made to answer the question?

9 A wagon is accelerating down a hill at constant acceleration. It took $1\,\text{s}$ more to accelerate from a velocity of $1\,\text{m}\,\text{s}^{-1}$ to a velocity of $5\,\text{m}\,\text{s}^{-1}$ than it took to accelerate from rest to a velocity of $1\,\text{m}\,\text{s}^{-1}$. Find the acceleration.

10 A driver sees a turning $100\,\text{m}$ ahead. She lets her car slow at constant deceleration of $0.4\,\text{m}\,\text{s}^{-2}$ and arrives at the turning $10\,\text{s}$ later. Find the velocity she is travelling at when she reaches the turning.

11 A cyclist is travelling at a velocity of $10\,\text{m}\,\text{s}^{-1}$ when he reaches the top of a slope, which is $80\,\text{m}$ long. There is a bend at the bottom of the slope, which it would be dangerous to go round any faster than $11\,\text{m}\,\text{s}^{-1}$. Because of gravity, if he did not pedal or brake he would accelerate down the slope at $0.1\,\text{m}\,\text{s}^{-2}$. To go as fast as possible but still reach the bottom at a safe speed should the cyclist brake, do nothing or pedal?

1.3 Equations of constant acceleration

In Worked example 1.5, you needed two equations to find the required answer. Wherever possible it is better to go directly from the information given to the required answer using just one equation because it is more efficient and reduces the number of equations to solve, and therefore reduces the likelihood of making mistakes.

There are five equations relating the five variables s, u, v, a and t. Each equation relates four of the five variables.

Two of these equations were introduced in Section 1.2, although the first one is normally given in the rearranged form shown in Key Point 1.8.

KEY POINT 1.8

For an object travelling with constant acceleration a, for time t, with initial velocity u, final velocity v and change of displacement s, we have

$$v = u + at$$

$$s = \tfrac{1}{2}(u + v)t$$

$$s = ut + \tfrac{1}{2}at^2$$

$$s = vt - \tfrac{1}{2}at^2$$

$$v^2 = u^2 + 2as$$

These equations are often referred to as the suvat equations.

TIP

In general, these equations are only valid if the acceleration is constant.

FAST FORWARD

In Chapter 6 you will consider how acceleration, speed, distance and time are related when the acceleration is not constant.

You will derive these equations in Exercise 1C.

WORKED EXAMPLE 1.6

a A go-kart travels down a slope of length 70 m. It is given a push and starts moving at an initial velocity of 3 m s^{-1} and accelerates at a constant rate of 2 m s^{-2}. Find its velocity at the bottom of the slope.

b Find the time taken for the go-kart to reach the bottom of the slope.

Answer

a The time, t, is unknown. | It is often useful to list what information is given and what is unknown.
The final velocity, v, is unknown.
$s = 70$
$u = 3$
$a = 2$

$v^2 = u^2 + 2as$ | Choose the equation with the known variables and the one required.
$= 3^2 + 2 \times 2 \times 70$
$= 289$ | In this case, we know u, a and s, and we want to find v.
$v = \pm 17$

We know that $v > 3$. | From the context, the velocity is increasing from 3 m s^{-1}, so only the positive solution is required.
$v = 17$ m s^{-1}

A negative velocity would indicate movement in the opposite direction.

b $s = ut + \frac{1}{2}at^2$ | Use a formula that involves given values rather than relying on your calculated values, as this will increase your chances of getting the correct answer even if your earlier answer was wrong.
$70 = 3t + \frac{1}{2} \times 2t^2$
$t^2 + 3t - 70 = 0$
$(t + 10)(t - 7) = 0$

$t = -10$ or $t = 7$ | Negative time would refer to time before the go-kart started its descent. Only the positive solution is required.
We know that $t > 0$ so $t = 7$ s.

Chapter 1: Velocity and acceleration

WORKED EXAMPLE 1.7

A trolley has a constant acceleration. After 2 s it has travelled 8 m and after another 2 s it has travelled a further 20 m. Find its acceleration.

Answer

Let the initial speed be v_1.	There are unknown velocities at three different times so simply using u and v may be insufficient and unclear.
Let the speed after 2 s be v_2.	
Let the speed after 4 s be v_3.	
Acceleration, a, is unknown.	List the information for the first 2 s.
$s = 8 \qquad t = 2$	There are too many unknowns to be able to calculate any of them at this stage.
$u = v_1 \qquad v = v_2$	
Acceleration, a, is unknown but the same as for the first 2 s.	List the information for the next 2 s.
$s = 20 \qquad t = 2$	Note that the speed after 2 s is the final speed for the first 2 s but the initial speed for the next 2 s so we can use the same letter to represent it.
$u = v_2 \qquad v = v_3$	
$s = vt - \tfrac{1}{2}at^2$	Since v_2 and a are the common unknowns in both stages of the motion, we will write equations relating them. First write the equation for the first stage of the motion.
$8 = 2v_2 - \tfrac{1}{2} \times a \times 2^2$	
$2v_2 - 2a = 8$	
$s = ut + \tfrac{1}{2}at^2$	We will again write the equation relating v_2 and a, but now for the second stage of the motion.
$20 = 2v_2 + \tfrac{1}{2} \times a \times 2^2$	
$2v_2 + 2a = 20$	
$4v_2 = 28$	Solve simultaneously by adding the equations and substituting the value for v_2 back in to one of the original equations.
$v_2 = 7 \text{ m s}^{-1}$	
So $a = 3 \text{ m s}^{-2}$.	
or	There is an alternative solution by considering the whole 4 s as one motion and creating equations involving v_1.
$s = ut + \tfrac{1}{2}at^2$	
$8 = 2v_1 + \tfrac{1}{2} \times a \times 2^2$	
$28 = 4v_1 + \tfrac{1}{2} \times a \times 4^2$	
giving $v_1 = 1 \text{ m s}^{-1}$.	
So $a = 3 \text{ m s}^{-2}$.	

EXERCISE 1C

1 For each part, assuming constant acceleration, write down the equation relating the four variables in the question and use it to find the missing variable.

   a Find s when $a = 3\,\text{m s}^{-2}$, $u = 2\,\text{m s}^{-1}$ and $t = 4\,\text{s}$.
   b Find s when $a = 2\,\text{m s}^{-2}$, $v = 17\,\text{m s}^{-1}$ and $t = 8\,\text{s}$.
   c Find a when $s = 40\,\text{m}$, $u = 3\,\text{m s}^{-1}$ and $t = 5\,\text{s}$.
   d Find a when $s = 28\,\text{m}$, $v = 13\,\text{m s}^{-1}$ and $t = 4\,\text{s}$.
   e Find a when $s = 24\,\text{m}$, $u = 2\,\text{m s}^{-1}$ and $v = 14\,\text{m s}^{-1}$.
   f Find u when $s = 45\,\text{m}$, $a = 1.5\,\text{m s}^{-2}$ and $t = 6\,\text{s}$.
   g Find v when $s = 24\,\text{m}$, $a = -2.5\,\text{m s}^{-2}$ and $t = 4\,\text{s}$.
   h Find s when $a = 0.75\,\text{m s}^{-2}$, $u = 2\,\text{m s}^{-1}$ and $v = 5\,\text{m s}^{-1}$.

2 Assuming constant acceleration, find the first time t, for positive t, at which the following situations occur.

   a Find t when $a = -2\,\text{m s}^{-2}$, $u = 10\,\text{m s}^{-1}$ and $s = 24\,\text{m}$.
   b Find t when $a = 0.5\,\text{m s}^{-2}$, $v = 5\,\text{m s}^{-1}$ and $s = 21\,\text{m}$.
   c Find t when $a = 1\,\text{m s}^{-2}$, $u = 3\,\text{m s}^{-1}$ and $s = 20\,\text{m}$.

3 Assuming constant acceleration, find v when $s = 6\,\text{m}$, $u = 5\,\text{m s}^{-1}$ and $a = -2\,\text{m s}^{-2}$ if the object has changed direction during the motion.

4 Assuming constant acceleration, find u when $s = 60\,\text{m}$, $v = 13\,\text{m s}^{-1}$ and $a = 1\,\text{m s}^{-2}$ if the object has not changed direction during the motion.

5 a Assuming constant acceleration, find v when $s = 18\,\text{m}$, $u = 3\,\text{m s}^{-1}$ and $a = 2\,\text{m s}^{-2}$.
   b Why is it not necessary to specify in this question whether the object has changed direction during the motion?

6 A car is travelling at a velocity of $20\,\text{m s}^{-1}$ when the driver sees the traffic lights ahead change to red. He decelerates at a constant rate of $4\,\text{m s}^{-2}$ and comes to a stop at the lights. Find how far away from the lights the driver started braking.

7 An aeroplane accelerates at a constant rate along a runway from rest until taking off at a velocity of $60\,\text{m s}^{-1}$. The runway is 400 m long. Find the acceleration of the aeroplane.

8 An aeroplane accelerates from rest along a runway at a constant rate of $4\,\text{m s}^{-2}$. It needs to reach a velocity of $80\,\text{m s}^{-1}$ to take off. Find how long the runway needs to be.

9 A motorcyclist sees that the traffic lights are red 40 m ahead of her. She is travelling at a velocity of $20\,\text{m s}^{-1}$ and comes to rest at the lights. Find the deceleration she experiences, assuming it is constant.

10 A driver sees the traffic lights change to red 240 m away when he is travelling at a velocity of $30\,\text{m s}^{-1}$. To avoid wasting fuel, he does not brake, but lets the car slow down naturally. The traffic lights change to green after 12 s, at the same time as the driver arrives at the lights.

   a Find the speed at which the driver goes past the lights.
   b What assumptions have been made to answer the question?

11 In a game of curling, competitors slide stones over the ice at a target 38 m away. A stone is released directly towards the target with velocity 4.8 m s^{-1} and decelerates at a constant rate of 0.3 m s^{-2}. Find how far from the target the stone comes to rest.

12 A golf ball is struck 10 m from a hole and is rolling towards the hole. It has an initial velocity of 2.4 m s^{-1} when struck and decelerates at a constant rate of 0.3 m s^{-2}. Does the ball reach the hole?

PS 13 A driverless car registers that the traffic lights change to amber 40 m ahead. The amber light is a 2 s warning before turning red. The car is travelling at 17 m s^{-1} and can accelerate at 4 m s^{-2} or brake safely at 8 m s^{-2}. What options does the car have?

P 14 The first two equations in Key Point 1.8 are $v = u + at$ and $s = \frac{1}{2}(u + v)t$. You can use these to derive the other equations.

  a By substituting for v in the second equation, derive $s = ut + \frac{1}{2}at^2$.

  b Derive the remaining two equations, $s = vt - \frac{1}{2}at^2$ and $v^2 = u^2 + 2as$, from the original two equations.

P 15 Show that an object accelerating with acceleration a from velocity u to velocity v, where $0 < u < v$, over a time t is travelling at a velocity of $\frac{u+v}{2}$ at time $\frac{1}{2}t$; that is, that at the time halfway through the motion the velocity of the object is the mean of the initial and final velocities.

P 16 Show that an object accelerating with acceleration a from velocity u to velocity v, where $0 < u < v$, over a displacement s is travelling at a speed of $\sqrt{\frac{v^2 + u^2}{2}}$ at a distance $\frac{1}{2}s$. Hence, prove that when the object does not change direction the speed at the midpoint of the distance is always greater than the mean of the initial and final speeds. Deduce also that the mean of the initial and final speeds occurs at a point closer to the start of the motion than the end.

1.4 Displacement–time graphs and multi-stage problems

It can be useful to show how the position of an object changes over time. You can do this using a displacement–time graph.

Imagine the following scenario. A girl is meeting a friend 1 km down a straight road. She takes 20 s to walk 30 m along the road to a bus stop. Then she waits 30 s for a bus, which takes her to a bus stop 20 m past her friend. The bus does not stop to pick anyone else up or drop them off. The journey takes 150 s. The girl takes 15 s to walk the 20 m back to meet her friend.

The graph would look like the one shown. You always show time on the x-axis and displacement on the y-axis. Notice you are defining the time as being measured from when the girl starts walking and the displacement from where she starts walking in the direction of her friend.

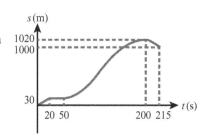

Where the graph is horizontal it indicates that the displacement is unchanged and therefore the girl is not moving. This was when she was waiting for the bus. If the graph is not horizontal it indicates the position is changing and the steepness of the line indicates how quickly it is changing.

Cambridge International AS & A Level Mathematics: Mechanics

A straight line on a displacement–time graph indicates a constant speed, as when the girl was walking to the bus stop. A curved line indicates a change in speed; for example, when the bus started moving after picking the girl up and when it slowed down to stop.

Notice that when the girl got off the bus to meet her friend she travelled in the opposite direction, so the change in her displacement and hence her velocity are negative. On the graph there is a negative gradient. The speed is the magnitude of the gradient, but the velocity includes the negative sign to indicate the direction.

Displacement–time graphs can have negative displacements below the x-axis, unlike distance–time graphs.

> **KEY POINT 1.9**
>
> The gradient of a displacement–time graph is equal to the velocity of the object.

> **FAST FORWARD**
>
> In Chapter 6, you will consider gradients of curved displacement–time graphs.

When sketching a graph of the motion of an object, you should show clearly the shape of the graph, and carefully distinguish a straight line from a curve. On a sketch you need to show only the key points. These include the intercept on the vertical axis, which is the initial position of the object, and any intercepts on the horizontal time axis, where the object is at the reference point. If there is more than one stage to the motion, you should clearly indicate the time and displacement of the object at the change in the motion.

WORKED EXAMPLE 1.8

A racing car passes the finish line of a race moving at a constant velocity of $60\,\mathrm{m\,s^{-1}}$. After $5\,\mathrm{s}$ it starts decelerating at $3\,\mathrm{m\,s^{-2}}$ until coming to rest. Sketch the displacement–time graph for the motion after the end of the race, measuring displacement from the finishing line.

Answer

Let t_1 be the time from the start of the first stage and t_2 be the time from the start of the second stage.

> The first stage of the journey is while the car travels at constant velocity.
>
> The second stage is while the car is decelerating.

Let s_1 be the displacement up to time t_1 during the first stage and s_2 the distance travelled during the second stage.

For the end of the first stage,

$s = vt$

So when $t_1 = 5$

$s_1 = 60 \times 5$

$ = 300\,\mathrm{m}$

> Find the displacement during the first stage because it will be marked on the sketch.

For the graph for $0 < t < 5$,

$s = 60t$

The graph of the first stage relates the variables s and t and is found using the equation for constant velocity.

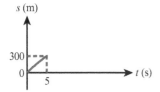

The first stage is a straight line graph with gradient 60 for 5 s.

The line starts at the origin because initial displacement is 0 m.

After 5 s the displacement is 300 m.

For the end of the second stage,

$v = u + at$

During the deceleration stage use an equation for constant acceleration to find the value of t_2 at the end of that stage of the motion.

So, $0 = 60 + (-3) \times t_2$

$t_2 = 20$

Use the final velocity from the first stage as the initial velocity for the second stage.

$v^2 = u^2 + 2as$

$0^2 = 60^2 + 2 \times (-3) \times s_2$

$s_2 = 600$

During the deceleration stage use an equation for constant acceleration to find the value of s_2 at the end of that stage of the motion.

Total time for the journey $= t_1 + t_2$

$= 5 + 20 = 25\,\text{s}$

The total time is the value to be marked on the sketch.

Total displacement $= s_1 + s_2$

$= 300 + 600$

$= 900\,\text{m}$

The total displacement from the finish line is the sum of both displacements.

For $5 < t < 25$,

$s = 300 + s_2$

and

$t = 5 + t_2$

The general displacement during the second stage is the sum of the displacement at the end of the first stage and the displacement during the second stage.

$s = ut + \dfrac{1}{2}at^2$

$s_2 = 60t_2 + \dfrac{1}{2} \times (-3) \times t_2^2$

$= 60t_2 - \dfrac{3}{2}t_2^2$

For the graph for $5 < t < 25$,

$s = 300 + 60t_2 - \dfrac{3}{2}t_2^2$

$= 300 + 60(t - 5) - \dfrac{3}{2}(t - 5)^2$

$= -\dfrac{3}{2}t^2 + 75t - \dfrac{75}{2}$

We can now find the equation of the curve in terms of s and t.

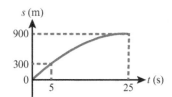

The graph for the second stage is a negative quadratic curve, valid for $5 < t < 25$ finishing horizontal at $t = 25$ (since $v = 0$).

The join between the graphs at $t = 5$ is smooth with the same gradient on the line before the join and the curve immediately after the join.

WORKED EXAMPLE 1.9

A cyclist is travelling at a velocity of $15\,\text{m}\,\text{s}^{-1}$ when he passes a junction. He then decelerates at a constant rate of $0.6\,\text{m}\,\text{s}^{-2}$ until coming to rest. A second cyclist travels at a constant velocity of $20\,\text{m}\,\text{s}^{-1}$ and passes the junction $4\,\text{s}$ after the first cyclist. Find the time at which the second cyclist passes the first and the displacement from the junction when that happens.

Answer

Let t_1 be the time from the first cyclist reaching the junction and t_2 be the time from the second cyclist reaching the junction.

> The cyclists pass the junction at different times, so it may be useful to define times for each of them separately.

Let s_1 be the displacement of the first cyclist from the junction and s_2 be the displacement of the second cyclist from the junction.

$s = ut + \dfrac{1}{2}at^2$

$s_1 = 15t_1 + \dfrac{1}{2} \times (-0.6)t_1^2$

> Find a formula for the displacement of the first cyclist.

$s = vt$

$s_2 = 20t_2$

> Find a formula for the displacement of the second cyclist from the junction, noting that the time is not measured from the same instant.

$t_2 = t_1 - 4$

So $s_2 = 20(t_1 - 4)$

> Find how the times are related.

$s_1 = s_2$

$\quad 15t_1 - 0.3t_1^2 = 20(t_1 - 4)$

$\quad 3t_1^2 + 50t_1 - 800 = 0$

$\quad (3t_1 + 80)(t_1 - 10) = 0$

$\quad t_1 = 10, -\dfrac{80}{3}$

So $t_1 = 10\,\text{s}$.

> One cyclist passes the other when they have the same displacement.

$$s_1 = 15 \times 10 + \frac{1}{2} \times (-0.6) \times 10^2$$
$$= 120\,\text{m}$$
Or $s_2 = 20(10 - 4)$
$$= 120\,\text{m}$$

Find the displacement from either formula as they should give the same answer.

WORKED EXAMPLE 1.10

Cyclist A is travelling at $16\,\text{m s}^{-1}$ when she sees cyclist B $15\,\text{m}$ ahead travelling at a constant velocity of $10\,\text{m s}^{-1}$. Cyclist A then slows at $1.5\,\text{m s}^{-2}$. Find the minimum gap between the cyclists.

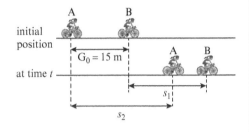

Answer

Let t be the time measured from when the cyclists are $15\,\text{m}$ apart and let the gap between the cyclists at time t be $G(t)$.

$G(t) = G_0 + s_1 - s_2$

$G(t) = 15 + 10t - (16t - 0.75t^2)$

$G(t) = 0.75(t - 4)^2 + 3$

Minimum gap is $3\,\text{m}$ at $4\,\text{s}$.

Or $16 - 1.5t = 10$

$t = 4$

$G(t) = 15 + 10t - (16t - 0.75t^2) = 3\,\text{m}$

Find the gap between the cyclists by adding the original gap and the change in displacement of the leading cyclist, and then subtracting the change of displacement of the following cyclist.

Complete the square to find the minimum gap and the time at which it occurs.

Alternatively, the closest distance is when the cyclists travel at the same speed because once the cyclist behind slows down the gap will increase again.

EXERCISE 1D

 1 Sketch the displacement–time graphs from the information given. In each case, consider north to be the positive direction and home to be the point from which displacement is measured.

   a Bob leaves his home and heads north at a constant speed of $3\,\text{m s}^{-1}$ for $10\,\text{s}$.

   b Jenny is $30\,\text{m}$ north of home and walks at a constant speed of $1.5\,\text{m s}^{-1}$ until reaching home.

   c Ryo is sitting still at a point $10\,\text{m}$ south of his home.

   d Nina is $300\,\text{m}$ north of her home. She drives south at a constant speed of $10\,\text{m s}^{-1}$, passing her home, until she has travelled a total of $500\,\text{m}$.

2 Sketch the displacement–time graphs from the information given. In each case consider upwards to be the positive direction and ground level to be the point from which displacement is measured. Remember to include the values for time and displacement at any points where the motion changes.

  a A firework takes off from ground level, accelerating upwards for 10 s with constant acceleration $4\,\text{m s}^{-2}$.

  b A ball is thrown upwards from a point 1 m above the ground with initial speed $5\,\text{m s}^{-1}$. It accelerates downwards at a constant rate of $10\,\text{m s}^{-2}$ until it stops moving upwards, when it is caught by someone standing on a ladder.

  c A rocket is falling at $10\,\text{m s}^{-1}$ at a height of 100 m above the ground when its engines turn on to provide a constant acceleration of $2\,\text{m s}^{-2}$ upwards. The engines remain on until the rocket has reached a height of 175 m above ground level.

  d A pebble is thrown upwards from the top of a cliff 18.75 m above the sea. It has initial speed $5\,\text{m s}^{-1}$. Initially it moves upwards, then it stops and falls downwards to reach the sea at the bottom of the cliff. Throughout the motion the pebble accelerates downwards at a constant rate of $10\,\text{m s}^{-2}$. Displacement is measured from the top of the cliff.

3 Sketch the displacement–time graphs from the information given. In each case consider forwards to be the positive direction and the traffic lights to be the point from which displacement is measured. Remember to include the values for time and displacement at any points where the motion changes.

  a A car is waiting at rest at the traffic lights. It accelerates at a constant rate of $3\,\text{m s}^{-2}$ for 5 s, then remains at constant speed for the next 10 s.

  b A motorbike passes the traffic lights at a constant speed of $10\,\text{m s}^{-1}$. After 6 s it starts to slow at a constant rate of $2\,\text{m s}^{-2}$ until it comes to rest.

  c A truck is moving at a constant speed of $8\,\text{m s}^{-1}$ and is approaching the traffic lights 60 m away. When it is 20 m away it accelerates at a constant rate of $2\,\text{m s}^{-2}$ to get past the lights before they change colour.

  d A scooter accelerates from rest 100 m before the traffic lights at a constant rate of $1.5\,\text{m s}^{-2}$ until it reaches $6\,\text{m s}^{-1}$. It then travels at this speed until it reaches a point 50 m beyond the traffic lights. At that point the scooter starts to slow at a constant rate of $1\,\text{m s}^{-2}$ until it stops.

4 The sketch shows a displacement–time graph of the position of a train passing a station. The displacement is measured from the entrance of the station to the front of the train. Find the equation of the displacement–time graph and hence the time at which the front of the train reaches the entrance of the station.

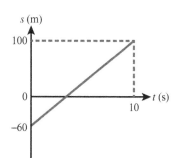

5 The sketch shows a displacement–time graph of a car slowing down with constant acceleration before coming to rest at a set of traffic lights.

  a The equation of the displacement–time graph can be written in the form $s = p(t - q)^2 + r$. Using the two points marked and the fact that the car is stationary at $t = 10$, find p, q and r.

  b By comparison with the equation $s = s_0 + ut + \frac{1}{2}at^2$, find the initial speed and acceleration of the car.

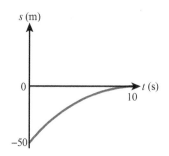

6 Two cars drive along the same highway. One car starts at junction 1, travelling north at a constant speed of 30 m s^{-1}. The second car starts at junction 2, which is 3 km north of junction 1, travelling south at a constant speed of 20 m s^{-1}.

   a Sketch the two displacement–time graphs on the same set of axes.

   b Find the equations of the two displacement–time graphs.

   c Solve the equations to find the time at which the cars pass each other and, hence, find the distance from junction 1 at which they pass.

7 Two trains travel on parallel tracks that are 5 km long. One starts at the southern end, travelling north at a constant speed of 25 m s^{-1}. The second train starts at the northern end 40 s later, travelling south at a constant speed of 15 m s^{-1}.

   a Sketch the two displacement–time graphs on the same set of axes.

   b Find the time for which the first train has been moving and the distance the first train has travelled when the trains pass each other.

8 A cyclist is stationary when a second cyclist passes, travelling at a constant speed of 8 m s^{-1}. The first cyclist then accelerates for 5 s at a constant rate of 2 m s^{-2} before continuing at constant speed until overtaking the second cyclist. By sketching both graphs, find the equations of the two straight-line sections of the graphs and, hence, find how long it is before the first cyclist overtakes the second.

9 Two rowing boats are completing a 2 km course. The first boat leaves, its crew rowing at a speed of 3.2 m s^{-1}. The second boat leaves some time later, its crew rowing at 4 m s^{-1}, and overtakes the first boat after the second has been travelling for 40 s.

   a Find how much earlier the second boat completes the course.

   b What assumption has been made in your answer?

10 The leader in a race has 500 m to go and is running at a constant speed of 4 m s^{-1}, but with 100 m to go increases her speed by a constant acceleration of 0.1 m s^{-2}. The second runner is 100 m behind the leader when the leader has 500 m to go, and running at 3.8 m s^{-1} when she starts to accelerate at a constant rate. Find the minimum acceleration she needs in order to win the race.

11 A van driver wants to pull out from rest onto a road where cars are moving at a constant speed of 20 m s^{-1}. When there is a large enough gap between cars, the van driver pulls out immediately after one car passes. She then accelerates at a constant rate of 4 m s^{-2} until moving at 20 m s^{-1}. To do this safely the car behind must always be at least 10 m away. Find the minimum length of the gap between the cars for the van driver to pull out.

12 A police motorcyclist is stationary when a car passes, driving dangerously at a constant speed of 40 m s^{-1}. At the instant the car passes, the motorcyclist gives chase, accelerating at 2.5 m s^{-2} until reaching a speed of 50 m s^{-1} before continuing at a constant speed. Show that the motorcyclist has not caught the car by the time he reaches top speed. Find how long after the car initially passed him the motorcyclist catches up to the car.

13 The front of a big wave is approaching a beach at a constant speed of 11.6 m s^{-1}. When it is 30 m away from a boy on the beach, the wave starts decelerating at a constant rate of 1.6 m s^{-2} and the boy walks away from the sea at a constant speed of 2 m s^{-1}. Show that the wave will not reach the boy and find the minimum distance between the boy and the wave.

14 Swimmers going down a waterslide 30 m long push themselves off with an initial speed of between 1 m s^{-1} and 2 m s^{-1}. They accelerate down with constant acceleration 0.8 m s^{-2} for the first 20 m before more water is added and the acceleration is 1 m s^{-2} for the last 10 m of the slide. For safety there must be at least 5 s between swimmers arriving at the bottom of the slide. Find the minimum whole number of seconds between swimmers being allowed to start the slide.

15 A ball is projected in the air with initial speed u and goes up and down with acceleration g downwards. A timer is at a height h. It records the time from the ball being projected until it passes the timer on the way up as t_1 and on the way down as t_2. Show that the total of the two times is independent of h and that the initial speed can be calculated as $u = \dfrac{g(t_1 + t_2)}{2}$. Show also that the difference between the times is given by $\dfrac{2\sqrt{u^2 - 2gh}}{g}$. Hence, find a formula for h in terms of t_1, t_2 and g.

1.5 Velocity–time graphs and multi-stage problems

As well as using a displacement–time graph, we can show the motion of an object on a velocity–time graph.

Imagine the following scenario. An athlete goes for a run. He starts at rest and gradually increases his speed over the first 30 s before maintaining the same speed of 5 m s^{-1} for 60 s. Then he gradually reduces his speed until coming to rest another 30 s later. The athlete then returns to his starting point by increasing his speed quickly at the start and continually trying to increase his speed for 90 s, but only managing to increase it by smaller and smaller amounts, peaking at 6 m s^{-1}. He then slows down over 10 s before coming to rest at his starting point.

The graph would look like the one shown here. You always show time on the x-axis and velocity on the y-axis.

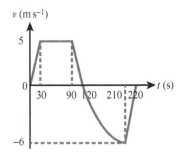

A horizontal graph line indicates that the velocity is unchanged and therefore the athlete is moving at constant speed. If the graph is not horizontal it indicates the velocity is changing and the steepness of the line indicates how quickly it is changing. Note that when the athlete returned to the start, the velocity became negative because the direction of motion changed.

KEY POINT 1.10

The gradient of a velocity–time graph is equal to the acceleration of the object.

From this graph you can see that the gradient is $\dfrac{v - u}{t}$, which is the same as the formula given for acceleration in Section 1.2.

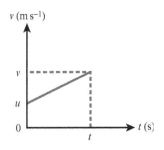

You can use the formula for the area of a trapezium to show that the area under the graph line is $\dfrac{1}{2}(u + v)t$, which is the same as the formula for the displacement. This rule can be generalised so that if the motion changes and the velocity–time graph has more than one line, the area under the graph may be found as the sum of separate areas under the lines.

Chapter 1: Velocity and acceleration

 KEY POINT 1.11

The area under the line of a velocity–time graph is the displacement of the object.

Note that in the previous scenario of the athlete, part of the graph is under the x-axis. The area below the axis is a 'negative area' and it indicates a negative displacement. In this particular example, the athlete started and ended at the same point and so the area above the axis should equal the area below the axis to indicate no overall change in displacement.

Note also that part of the graph is curved. This indicates that the acceleration is not constant.

 FAST FORWARD

In Chapter 6 you will consider gradients of and areas under curved velocity–time graphs.

MODELLING ASSUMPTIONS

In the same way as we asked if it is reasonable to assume constant speed, we might ask if it is reasonable to assume constant acceleration. In many cases it is close enough, but it is often harder to maintain the same acceleration when moving at high speeds.

In scenarios involving people, we often say that someone initially is not moving and then walks at a given speed. We assume that the change is instantaneous. In the case of walking at low speeds, the time taken to reach that speed is sufficiently small that it is not a bad assumption, but for runners there may be some error in making that assumption.

 DID YOU KNOW?

Olympic sprinters take about 60 m to reach top speed. By the end of the 100 m race they are normally starting to slow down. You might expect that, because runners start to slow down after about 100 m, race times for 200 m will be more than double the times for 100 m. In fact, for most of the time since world records were recorded, the 200 m world record has been less than double the 100 m record because the effect of starting from a stationary position is larger than the effect of slowing down by a small amount for the second 100 m.

WORKED EXAMPLE 1.11

a Arthur travels at a constant speed of $5\,\text{m}\,\text{s}^{-1}$ for 10 s and then decelerates at a constant rate of $0.5\,\text{m}\,\text{s}^{-2}$ until coming to rest. Sketch the velocity–time graph for his motion.

b Brendan travels at a constant $4\,\text{m}\,\text{s}^{-1}$ starting from the same time and place. Show that Arthur and Brendan are travelling at the same speed after 12 s and, hence, find the furthest Arthur gets ahead of Brendan.

c Show that for $t > 10$ the gap between them is given by $g(t) = -\dfrac{1}{4}t^2 + 6t - 25$ and, hence, find the time when Brendan overtakes Arthur.

Cambridge International AS & A Level Mathematics: Mechanics

Answer

a Let T be the time spent in deceleration. ⋯⋯⋯⋯⋯ Use an equation of constant acceleration to
$v = u + at$ find the time for the second stage of the motion.
$0 = 5 - 0.5T$
$T = 10\,\text{s}$
$t = T + 10$
$= 20\,\text{s}$

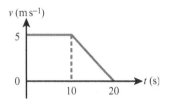

⋯⋯⋯⋯⋯⋯⋯⋯⋯⋯⋯⋯⋯⋯⋯⋯⋯ Constant velocity means a horizontal line.

Deceleration from a positive velocity means a negative gradient.

The x-axis intercept is at the total time from the start.

b $v\,(\text{m s}^{-1})$

⋯⋯⋯⋯⋯⋯⋯⋯⋯⋯⋯⋯⋯⋯⋯⋯⋯ Find where the lines cross to solve when velocities are equal.

$v = 5 - 0.5(t - 10) = 4$ ⋯⋯⋯⋯⋯⋯⋯⋯⋯⋯⋯⋯ Time in the second stage is 10s less than total
$t = 12\,\text{s}$ time since start.

$s_A = 5 \times 10 + \dfrac{1}{2} \times (5 + 4) \times 2 = 59\,\text{m}$ ⋯⋯⋯⋯⋯⋯ The largest gap between them is equal to the
difference in displacements at the time when
$s_B = 4 \times 12 = 48\,\text{m}$ they have the same velocity. After this time

Therefore, the largest gap is 11m. Brendan is travelling faster than Arthur and so starts to catch up.

Distance travelled is area under graph: a rectangle plus a trapezium for Arthur and a rectangle for Brendan.

c Starting gap = 0 ⋯⋯⋯⋯⋯⋯⋯⋯⋯⋯⋯⋯⋯⋯⋯⋯ The gap at time t is the starting gap plus the
At time t: distance covered by the leading person minus
the distance covered by the following person.
$s_A = 50 + \dfrac{1}{2}(t - 10)\left(5 + \left(5 - \dfrac{1}{2}(t - 10)\right)\right)$

$= 50 + 5(t - 10) - \dfrac{1}{4}(t - 10)^2$ For $t > 10$, Arthur is in the second stage of the motion, so the total distance is the distance covered in the first stage plus the distance

$s_B = 4t$ covered in the second stage up to time t.

$g(t) = 0 + 50 + 5(t - 10) - \dfrac{1}{4}(t - 10)^2 - 4t$

$= -\dfrac{1}{4}t^2 + 6t - 25$

Brendan overtakes when the gap is 0, so ..... Solve $g(t) = 0$ to find t.

$-\dfrac{1}{4}t^2 + 6t - 25 = 0$

$t^2 - 24t + 100 = 0$

$t = 5.37$ or 18.6

Since the equations are valid only for $t > 10$, ... Check the context and validity of the equations to determine which is the relevant solution.
$t = 18.6\,\text{s}$.

EXERCISE 1E

1 Sketch the velocity–time graphs from the information given. In each case take north to be the positive direction.

   a Rinesh starts from rest, moving north with a constant acceleration of $3\,\text{m s}^{-2}$ for $5\,\text{s}$.

   b Wendi is moving north at $2\,\text{m s}^{-1}$ when she starts to accelerate at a constant rate of $0.5\,\text{m s}^{-2}$ for $6\,\text{s}$.

   c Dylon is moving south at a constant speed of $4\,\text{m s}^{-1}$.

   d Susan is moving north at $6\,\text{m s}^{-1}$ when she starts decelerating at a constant rate of $0.3\,\text{m s}^{-2}$ until she comes to rest.

2 Sketch the velocity–time graphs from the information given. In each case take upwards to be the positive direction.

   a A ball is thrown up in the air from the surface of a pond with initial velocity $20\,\text{m s}^{-1}$. It accelerates downwards under gravity with constant acceleration $10\,\text{m s}^{-2}$. Once it has reached its highest point it falls until it hits the surface of the pond and goes underwater. Under the water it continues to accelerate with constant acceleration $1\,\text{m s}^{-2}$ for $1\,\text{s}$.

   b A parachutist falls from a helicopter that is flying at a constant height. She accelerates downwards at a constant rate of $10\,\text{m s}^{-2}$ for $0.5\,\text{s}$ before the parachute opens. She then remains at constant speed for $5\,\text{s}$.

   c A hot-air balloon is floating at a constant height before descending to a lower height. It descends with constant acceleration $5\,\text{m s}^{-2}$ for $6\,\text{s}$, then the burner is turned on and the balloon decelerates at a constant rate of $2\,\text{m s}^{-2}$ until it is no longer descending.

   d A firework takes off from rest and accelerates upwards for $7\,\text{s}$ with constant acceleration $5\,\text{m s}^{-2}$, before decelerating at a constant rate of $10\,\text{m s}^{-2}$ until it explodes at the highest point of its trajectory.

3 The graph shows the motion of a motorcyclist when he starts travelling along a highway until reaching top speed. Find the distance covered in reaching that speed.

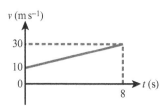

4 The graph shows the motion of a ball when it is thrown upwards in the air until it hits the ground. Find the height above the ground from which it was thrown.

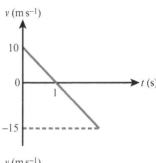

5 The sketch graph shows the motion of a boat. Find the distance the boat travels during the motion.

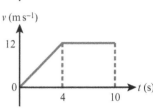

6 The graph shows the journey of a cyclist travelling in a straight line from home to school. Find the distance between her home and the school.

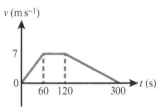

7 A racing car is being tested along a straight 1 km course. It starts from rest, accelerating at a constant rate of $10\,\mathrm{m\,s^{-2}}$ for 5 s. It then travels at a constant speed until a time t s after it started moving. Show that the distance covered by time t is given by $s = 125 + 50(t - 5)$. Hence, find how long it takes to complete the course.

8 A rowing boat accelerates from rest at a constant rate of $0.4\,\mathrm{m\,s^{-2}}$ for 5 s. It continues at constant velocity for some time until decelerating to rest at a constant rate of $0.8\,\mathrm{m\,s^{-2}}$. In total, the boat covers a distance of 30 m. Find how long was spent at constant speed.

9 A cyclist accelerates at a constant rate for 10 s, starting from rest and reaching a speed of $v\,\mathrm{m\,s^{-1}}$. She then remains at that speed for a further 20 s. At the end of this she has travelled 300 m in total. Find the value of v.

10 A boat accelerates from rest at a rate of $0.2\,\mathrm{m\,s^{-2}}$ to a speed $v\,\mathrm{m\,s^{-1}}$. It then remains at that speed for a further 30 s. At the end of this it has travelled 400 m in total.

   a Find the value of v.

   b What assumptions have been made to answer the question?

11 A crane lifts a block from ground level at a constant speed of $v\,\mathrm{m\,s^{-1}}$. After 5 s the block slips from its shackles and decelerates at $10\,\mathrm{m\,s^{-2}}$. It reaches a maximum height of 6 m. Find the value of v.

Chapter 1: Velocity and acceleration

12 A car is at rest when it accelerates at $5\,\text{m}\,\text{s}^{-2}$ for $4\,\text{s}$. It then continues at a constant velocity. At the instant the car starts moving, a truck passes it, moving at a constant speed of $22\,\text{m}\,\text{s}^{-1}$. After $10\,\text{s}$ the truck starts slowing at $1\,\text{m}\,\text{s}^{-2}$ until coming to rest.

   a Show that the velocities are equal after $12\,\text{s}$ and, hence, find the maximum distance between the car and the truck.

   b Show that the distance covered at a time $t\,\text{s}$ from the start by the car and the truck, for $t > 10$, are given by $40 + 20(t - 4)$ and $220 + 22(t - 10) - \dfrac{1}{2}(t - 10)^2$, respectively. Hence, find the time at which the car passes the truck.

13 Two cyclists are having a race along a straight road. Bradley starts $50\,\text{m}$ ahead of Chris. Bradley starts from rest, accelerates to $15\,\text{m}\,\text{s}^{-1}$ in $10\,\text{s}$ and remains at this speed for $40\,\text{s}$ before decelerating at $0.5\,\text{m}\,\text{s}^{-2}$. Chris starts $5\,\text{s}$ later than Bradley. He starts from rest, accelerates to $16\,\text{m}\,\text{s}^{-1}$ in $8\,\text{s}$ and maintains this speed.

   a Show that Bradley is still ahead when he starts to slow down, and find how far ahead he is.

   b Find the amount of time Bradley has been cycling when he is overtaken by Chris.

PS 14 A driver travelling at $26\,\text{m}\,\text{s}^{-1}$ sees a red traffic light ahead and starts to slow at $3\,\text{m}\,\text{s}^{-2}$ by removing her foot from the accelerator pedal. A little later she brakes at $5\,\text{m}\,\text{s}^{-2}$ and comes to rest at the lights after $6\,\text{s}$.

   a Sketch the velocity–time graph of the motion.

   b Find the equations of the two sections of the graph.

   c Hence, find the time when the driver needs to start braking.

P 15 A car accelerates from rest to a speed $v\,\text{m}\,\text{s}^{-1}$ at a constant acceleration. It then immediately decelerates at a constant deceleration until coming back to rest $t\,\text{s}$ after starting the motion.

   a Show that the distance travelled is independent of the values of the acceleration and deceleration.

   b Suppose instead the car spends a time $T\,\text{s}$ at speed $v\,\text{m}\,\text{s}^{-1}$ but still returns to rest after a total of $t\,\text{s}$ after starting the motion. Show that the distance travelled is independent of the values of the acceleration and deceleration.

1.6 Graphs with discontinuities

What happens when a ball bounces or is struck by a bat? It would appear that the velocity instantaneously changes from one value directly to a different value. If this did happen instantaneously, the acceleration would be infinite. In practice, the change in velocity happens over a tiny amount of time that it is reasonable to ignore, so we will assume the change is instantaneous.

The velocity–time graph will have a discontinuity, as shown in the following graph, as the velocity instantaneously changes.

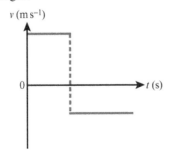

The displacement–time graph cannot have a discontinuity, but the gradient will instantaneously change, so the graph will no longer be smooth at the join between two stages of the motion. For the velocity–time graph shown, the displacement–time graph will look like the following.

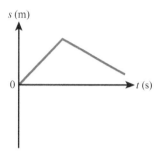

KEY POINT 1.12

On the velocity–time graph of an object that instantaneously changes velocity by bouncing or being struck, the change is represented by a vertical dotted line from the velocity before impact to the velocity after impact.

MODELLING ASSUMPTIONS

In practice, the objects may not instantaneously change velocity. In the example of a tennis ball being hit by a racket, the strings stretch very slightly and spring back into shape. It is during this time that the ball changes velocity. In the case of a tennis ball striking a solid wall or a solid object striking the ball, the ball may compress slightly during contact before springing back into shape. In these cases, the time required to change is so small that you can ignore it. By modelling the objects as particles, you can assume the objects do not lose shape and the time in contact is sufficiently small to be negligible.

DID YOU KNOW?

Golf balls look and feel solid, but in the instant after impact from a golf club moving at around 200 km h^{-1} the ball appears to squash so that its length is only about 80% of its original diameter and its width increases.

WORKED EXAMPLE 1.12

a A ball is travelling at a constant speed of 10 m s^{-1} for 2 s until it strikes a wall. It bounces off the wall at 5 m s^{-1} and maintains that speed until it reaches where it started. When it passes that point it decelerates at 1 m s^{-2}. Find the times and displacements when each change in the motion occurs.

b Sketch a velocity–time graph and a displacement–time graph for the motion. Measure displacements as distances from the starting point and the original direction of motion as positive.

Chapter 1: Velocity and acceleration

Answer

a The distance to the wall is
$s = 10 \times 2 = 20$ m

The time between hitting the wall and returning to the starting point is, therefore,
$\frac{20}{5} = 4$ s so $t = 6$ s

> Note that times are measured from the start of the motion.

The time from starting to decelerate until it stops is
$\frac{0 - (-5)}{1} = 5$ s so $t = 11$ s

The distance covered is
$5 \times 5 + \frac{1}{2} \times (-1) \times 5^2 = 12.5$ m
so displacement is
$s = -12.5$ m

> Note that displacements are measured from the starting position, taking the original direction as positive.
>
> Although decelerating, the acceleration is positive because the velocity is negative.

b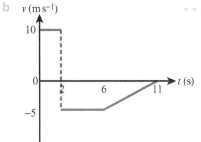

> Notice the graph is discontinuous at $t = 2$.
>
> Although the ball is decelerating after $t = 6$, the gradient is positive because the velocity is negative.

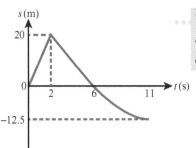

> Notice that at $t = 2$ the gradient is different on either side of the cusp. This indicates a discontinuity in the velocity.

EXERCISE 1F

1 An ice hockey puck slides along a rink at a constant speed of 10 m s^{-1}. It strikes the boards at the edge of the rink 20 m away and slides back along the rink at 8 m s^{-1} until going into the goal 40 m from the board. Sketch a velocity–time graph and a displacement–time graph for the motion, measuring displacement from the starting point in the original direction of motion.

2 A bowling ball rolls down an alley with initial speed 8 m s^{-1} and decelerates at a constant rate of 0.8 m s^{-2}. After 2.5 s it strikes a pin and instantly slows down to 2 m s^{-1}. It continues to decelerate at the same constant rate until coming to rest. Sketch a velocity–time graph and a displacement–time graph for the motion.

3 In a game of blind cricket, a ball is rolled towards a player with a bat 20 m away, who tries to hit the ball. On one occasion, the ball is rolled towards the batsman at a constant speed of $4\,\text{m s}^{-1}$. The batsman hits the ball back directly where it came from with initial speed $6\,\text{m s}^{-1}$ and decelerating at a constant rate of $0.5\,\text{m s}^{-2}$. Sketch a velocity–time graph and a displacement–time graph for the motion, taking the original starting point as the origin and the original direction of motion as positive.

4 A ball is dropped from rest 20 m above the ground. It accelerates towards the ground at a constant rate of $10\,\text{m s}^{-2}$. It bounces on the ground and leaves with a speed that is half the speed it struck the ground originally. The ball is then caught when it reaches the highest point of its bounce. Sketch a velocity–time graph and a displacement–time graph for the motion, measuring displacement above the ground.

5 A ball is thrown towards a wall, which is 5 m away, at $2.25\,\text{m s}^{-1}$. It slows down at a constant rate of $0.2\,\text{m s}^{-2}$ until it strikes the wall. It bounces back at 80% of the speed it struck the wall originally. It again slows down at a constant rate of $0.2\,\text{m s}^{-2}$ until coming to rest. Sketch a velocity–time graph and a displacement–time graph for the motion, measuring displacement from the wall, taking the direction away from the wall as positive.

6 A billiard ball is on the centre spot of a 6 m long table and is struck towards one of the cushions with initial speed $3.1\,\text{m s}^{-1}$. It slows on the table at $0.2\,\text{m s}^{-2}$. When it bounces off the cushion its speed reduces to 70% of the speed with which it struck the cushion. The ball is left until it comes to rest.

   a Sketch the velocity–time and displacement–time graphs for the ball, taking the centre of the table as the origin for displacement and the original direction of motion as positive.

   b What assumptions have been made in your answer?

7 A ball is released from rest 20 m above the ground and accelerates under gravity at $10\,\text{m s}^{-2}$. When it bounces its speed halves. If bounce n occurs at time t_n the speed after the bounce is v_n. Show that $v_n = 15 - 2.5t_n$ and deduce that, despite infinitely many bounces, the ball stops bouncing after 6 s.

Checklist of learning and understanding

- The equations of constant acceleration are:

  $v = u + at$

  $s = \frac{1}{2}(u + v)t$

  $s = ut + \frac{1}{2}at^2$

  $s = vt - \frac{1}{2}at^2$

  $v^2 = u^2 + 2as$

- A displacement–time graph shows the position of an object at different times. The gradient is equal to the velocity.
- A velocity–time graph shows how quickly an object is moving at a given time. The gradient is equal to the acceleration. The area under the graph is equal to the displacement.

Chapter 1: Velocity and acceleration

END-OF-CHAPTER REVIEW EXERCISE 1

1. A man and his young son play a game. The man rolls a ball along the ground. His son runs after the ball to fetch it.

   a The ball starts rolling at $10\,\text{m s}^{-1}$ but decelerates at a constant rate of $2\,\text{m s}^{-2}$. Find the distance covered when it comes to rest.

   b Once the ball has stopped, the boy runs to fetch it. He starts from rest beside his father and accelerates at a constant rate of $2\,\text{m s}^{-2}$ for $3\,\text{s}$ before maintaining a constant speed. Find the time taken to reach the ball.

2. A car is travelling at $15\,\text{m s}^{-1}$ when the speed limit increases and the car accelerates at a constant rate of $3\,\text{m s}^{-2}$ until reaching a top speed of $30\,\text{m s}^{-1}$.

   a Find the distance covered until reaching top speed.

   b Once the car is at top speed, there is a set of traffic lights $600\,\text{m}$ away. The car maintains $30\,\text{m s}^{-1}$ until it starts to decelerate at a constant rate of $5\,\text{m s}^{-2}$ to come to rest at the lights. Find the time taken from reaching top speed until it comes to rest at the traffic lights.

3. In a race, the lead runner is $60\,\text{m}$ ahead of the chaser with $200\,\text{m}$ to go and is running at $4\,\text{m s}^{-1}$. The chaser is running at $5\,\text{m s}^{-1}$.

   a Find the minimum constant acceleration required by the chaser to catch the lead runner.

   b If the lead runner is actually accelerating at a constant rate of $0.05\,\text{m s}^{-2}$, find the minimum constant acceleration required by the chaser to catch the lead runner.

4. A jet aeroplane coming in to land at $100\,\text{m s}^{-1}$ needs $800\,\text{m}$ of runway.

   a Find the deceleration, assumed constant, the aeroplane can produce.

   b On an aircraft carrier, the aeroplane has only $150\,\text{m}$ to stop. There are hooks on the aeroplane that catch arresting wires to slow it down. If the aeroplane catches the hook $50\,\text{m}$ after landing, find the deceleration during the last $100\,\text{m}$.

5. The sketch shows a velocity–time graph for a sled going down a slope. Sketch the displacement–time graph, marking the displacements at each change in the motion.

   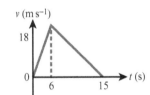

6. The sketch shows a velocity–time graph for rowers in a race. Given that the race is $350\,\text{m}$ long and finishes at time $50\,\text{s}$, find the value of v.

   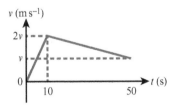

7 A footballer kicks a ball directly towards a wall 10 m away and walks after the ball in the same direction at a constant $2\,\text{m s}^{-1}$. The ball starts at $4\,\text{m s}^{-1}$ but decelerates at a constant rate of $0.5\,\text{m s}^{-2}$. When it hits the wall it rebounds to travel away from the wall at the same speed with which it hit the wall.

  a Find the time after the initial kick when the ball returns to the footballer.

  b What assumptions have been made in your answer?

8 An entrant enters a model car into a race. The car accelerates from rest at a constant rate of $2\,\text{m s}^{-2}$ down a slope. When it crosses the finishing line a firework is set off at the end of the course. The sound travels at $340\,\text{m s}^{-1}$. The time between the entrant starting and the firework being heard at the start of the course is $12\,\text{s}$.

  a Find the length of the course.

  b Find the actual time it took for the model car to complete the course.

9 A lion is watching a zebra from 35 m behind it. Both are stationary. The lion then starts chasing by accelerating at a constant rate of $3\,\text{m s}^{-2}$ for $5\,\text{s}$. Once at top speed the lion maintains that speed for $3\,\text{s}$ before decelerating at $0.5\,\text{m s}^{-2}$. The zebra starts moving 1s after the lion started, accelerating at a constant rate of $2\,\text{m s}^{-2}$ for $7\,\text{s}$ before maintaining a constant speed.

  a Show that the lion has not caught the zebra after $8\,\text{s}$.

  b Show that the gap between them at time t s, for $t > 8$, after the start of the lion's motion is given by $\frac{1}{4}t^2 - 5t + \frac{51}{2}$ and, hence, determine when the lion catches the zebra, or when the lion gets closest and how close it gets.

10 A car is behind a tractor on a single-lane straight road with one lane in each direction. Both are moving at $15\,\text{m s}^{-1}$. The speed limit is $25\,\text{m s}^{-1}$, so the car wants to overtake. The safe distance between the car and the tractor is 20 m.

  a To overtake, the car goes onto the other side of the road and accelerates at a constant rate of $2\,\text{m s}^{-2}$ until reaching the speed limit, when it continues at constant speed. Show that the distance the car is ahead of the tractor at time t s after it starts to accelerate is given by $t^2 - 20$ for $0 < t \leq 5$, and deduce that the car is not a safe distance ahead of the tractor before reaching the speed limit.

  b The car pulls in ahead of the tractor once it is a safe distance ahead. Find the total time taken from the start of the overtaking manoeuvre until the car has safely overtaken the tractor.

  c To overtake safely on the single-lane road, when the car returns to the correct side of the road in front of the tractor there must be a gap between the car and oncoming traffic of at least 20 m. Assuming a car travelling in the opposite direction is moving at the speed limit, find the minimum distance it must be from the initial position of the overtaking car at the point at which it starts to overtake.

11 Two hockey players are practising their shots. They are 90 m apart and hit their balls on the ground directly towards each other. The first player hits his ball at $6\,\text{m s}^{-1}$ and the other hits hers at $4\,\text{m s}^{-1}$. Both balls decelerate at $0.1\,\text{m s}^{-2}$. Find the distance from the first player when the balls collide.

12 The sketch shows a velocity–time graph for a skier going down a slope. Given that the skier covers 80 m during the first stage of acceleration, find the total distance covered.

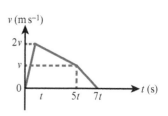

13 Two trains are travelling towards each other, one heading north at a constant speed of u m s^{-1} and the other heading south at a constant speed of v m s^{-1}. When the trains are a distance d m apart, a fly leaves the northbound train at a constant speed of w m s^{-1}. As soon as it reaches the other train, it instantly turns back travelling at w m s^{-1} in the other direction. Show that the fly meets the southbound train having travelled a distance of $\dfrac{wd}{w+v}$ and returns to the northbound train when the train has travelled a distance of $\dfrac{2uwd}{(w+v)(w+u)}$.

14 Two cars are on the same straight road, the first one s m ahead of the second and travelling in the same direction. The first car is moving at initial speed v m s^{-1} away from the second car. The second car is moving at initial speed u m s^{-1}, where $u > v$. Both cars decelerate at a constant rate of a m s^{-2}.

   a Show that the second car overtakes at time $t = \dfrac{s}{u-v}$ irrespective of the deceleration, provided the cars do not come to rest before the second one passes.

   b Show also that the distance from the starting point of the second car to the point where it overtakes depends on a and find a formula for that distance.

15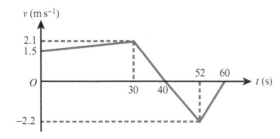

A woman walks in a straight line. The woman's velocity t seconds after passing through a fixed point A on the line is v m s^{-1}. The graph of v against t consists of 4 straight line segments (see diagram).

The woman is at the point B when $t = 60$. Find

   i the woman's acceleration for $0 < t < 30$ and for $30 < t < 40$, [3]

   ii the distance AB, [2]

   iii the total distance walked by the woman. [1]

Cambridge International AS & A Level Mathematics 9709 Paper 43 Q1 November 2011

16 A car travels in a straight line from A to B, a distance of 12 km, taking 552 seconds. The car starts from rest at A and accelerates for T_1 s at 0.3 m s^{-2}, reaching a speed of V m s^{-1}. The car then continues to move at V m s^{-1} for T_2 s. It then decelerates for T_3 s at 1 m s^{-2}, coming to rest at B.

   i Sketch the velocity–time graph for the motion and express T_1 and T_3 in terms of V. [3]

   ii Express the total distance travelled in terms of V and show that $13V^2 - 3312V + 72\,000 = 0$. Hence find the value of V. [5]

Cambridge International AS & A Level Mathematics 9709 Paper 43 Q5 November 2013

17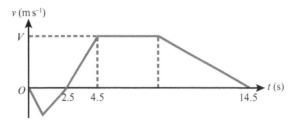

The diagram shows the velocity–time graph for a particle P which travels on a straight line AB, where $v\,\mathrm{m\,s^{-1}}$ is the velocity of P at time t s. The graph consists of five straight line segments. The particle starts from rest when $t = 0$ at a point X on the line between A and B and moves towards A. The particle comes to rest at A when $t = 2.5$.

i Given that the distance XA is 4 m, find the greatest speed reached by P during this stage of the motion. [2]

In the second stage, P starts from rest at A when $t = 2.5$ and moves towards B. The distance AB is 48 m. The particle takes 12 s to travel from A to B and comes to rest at B. For the first 2 s of this stage P accelerates at $3\,\mathrm{m\,s^{-2}}$, reaching a velocity of $V\,\mathrm{m\,s^{-1}}$. Find

ii the value of V, [2]

iii the value of t at which P starts to decelerate during this stage, [3]

iv the deceleration of P immediately before it reaches B. [2]

Cambridge International AS & A Level Mathematics 9709 Paper 42 Q6 November 2010

Chapter 2
Force and motion in one dimension

In this chapter you will learn how to:

- relate force to acceleration
- use combinations of forces to calculate their effect on an object
- include the force on an object due to gravity in a force diagram and calculations
- include the normal contact force on a force diagram and in calculations.

Cambridge International AS & A Level Mathematics: Mechanics

PREREQUISITE KNOWLEDGE		
Where it comes from	What you should be able to do	Check your skills
Chapter 1	Use equations of constant acceleration.	1 Find t when $u = 4$, $v = 6$ and $a = 5$. 2 Find v when $u = 3$, $a = 4$ and $s = 2$.

What is a force and how does it affect motion?

When an object is moving, how will it continue to move? What makes it speed up or slow down? What affects the acceleration or deceleration of an object? These are questions that have been considered by many philosophers over the course of history. The first person to publish what is now considered to be the correct philosophy was Isaac Newton (1643–1727), which is why Mechanics is often referred to as Newtonian Mechanics. Newton used his theories to explain and accurately predict the movements of planets and the Moon, as well as to explain tides and the shape of the Earth.

Newton was not the first to consider the idea of a force. Aristotle (384–322 BCE), the ancient Greek philosopher, believed that a cause (or force) was the thing that had an effect on an object to make it move. He came up with a theory of how force was related to motion, but he did not define forces clearly and he had no supporting evidence for his claims. His theory led to the conclusion that heavier bodies fall to Earth faster than lighter bodies, which is not true.

2.1 Newton's first law and relation between force and acceleration

A **force** is something that can cause a change in the motion of an object. There are many different types of force that can act on an object. Think of some ways that an object on a table can be made to either start moving or change its speed.

Most forces are caused by objects being in contact with other objects. Pushing or dragging an object can cause it to accelerate. A force can act through a string under **tension**; pulling on the string can cause an attached object to speed up or slow down. A solid rod can act like a string, but it can also push back when it is under **compression** and there is a **thrust** in the rod.

Objects that are in contact with each other may experience **friction**, which may include air **resistance**. This generally slows objects down, but sometimes friction is required to cause the motion. For example, a car engine gets the car to move by using the friction between the tyres and the road to start the wheels rolling along the ground. In icy or muddy conditions there is not much friction so cars cannot accelerate quickly.

If you gently push an object off the edge of a table it will accelerate towards the ground despite nothing (apart from the air) being in contact with it. This is because there is a force due to **gravity**. In fact, there is a force due to gravity between all objects, but other than the gravity pulling objects towards the Earth, the forces are so small as to be negligible.

Gravity is the only force you will consider in this course that acts on an object without being in contact with the object. Other forces that act in this way; for example, magnetic attraction or repulsion, will not be considered in this course.

Newton progressed from using mathematics to calculate the position, speed and acceleration of an object with constant acceleration to explaining why objects move as they do. He produced three laws of motion that are still used today in many situations to calculate and describe how objects move.

DID YOU KNOW?

Isaac Newton published his work in a book called *Philosophiae Naturalis Principia Mathematica* in 1687, often known simply as the *Principia*.

Chapter 2: Force and motion in one dimension

 KEY POINT 2.1

Newton's first law states that an object remains at rest or continues to move at a constant velocity unless acted upon by a net force.

This is not immediately obvious because the forces are not visible. You see that objects sliding along the ground slow down and eventually stop without anyone trying to slow them down and without the object hitting another object. A ball moving through the air appears to be changing direction as it falls under gravity, yet it does not touch anything while changing direction.

Newton's second law expresses how a force relates to the motion of an object. For an object of constant mass, the net force acting on the object is proportional to the product of its mass and acceleration.

$$F \propto ma$$

The force is measured in newtons (symbol N). One newton is defined as the amount of force required to accelerate $1\,\text{kg}$ at $1\,\text{m}\,\text{s}^{-2}$. Using kilograms for the unit of mass, metres for length and seconds for time, so that acceleration is in $\text{m}\,\text{s}^{-2}$, the constant of proportionality is 1.

 KEY POINT 2.2

Newton's second law leads to the equation

$$\text{Force} = \text{mass} \times \text{acceleration}$$

Force is a vector quantity, so may be positive or negative depending on which direction is assigned to be positive.

MODELLING ASSUMPTIONS

In all cases at this stage you will consider objects as particles, so you can ignore any complexities due to the shape of the object. For many of the problems, it will have such a small effect that you can treat it as negligible. This means that the error it causes in the calculations is small enough to be ignored.

For example, when you consider an object like a bag of sand, if the mass of the bag is sufficiently small compared to the mass of the sand you say it is negligible and the bag is termed light.

In some questions, there is a general force called resistance, which acts in the opposite direction to motion. This may be due to friction, air resistance or both. In some situations you will ignore resistance forces altogether. This is a modelling assumption that you make to simplify the situation.

 DID YOU KNOW?

Aristotle thought that at every moment something must be causing an object to continue to move, so an object flying through the air must be pushed by the air to continue moving. Newton was the first to contradict him.

 FAST FORWARD

Newton's third law relates forces between objects and their effects on each other. You will learn about this in Chapter 5.

 FAST FORWARD

Friction will be considered in more detail in Chapter 4.

Cambridge International AS & A Level Mathematics: Mechanics

WORKED EXAMPLE 2.1

a A cyclist and his bike have a combined mass of 100 kg. The cyclist accelerates from rest with acceleration $0.2 \, \text{m s}^{-2}$. Find the force the cyclist generates.

b A second cyclist and his bike have a combined mass of 80 kg. This cyclist accelerates from rest by generating the same force as the first cyclist. Find her acceleration.

c The first cyclist is travelling at $20 \, \text{m s}^{-1}$ when he starts to brake with a force of 500 N. Find the distance covered while coming to rest.

Answer

a $F = ma = 100 \times 0.2 = 20 \, \text{N}$ .... Use $F = ma$.

b $F = ma$ .... Use $F = ma$ and solve the equation.
$20 = 80a$
$a = \dfrac{1}{4}$ or $0.25 \, \text{m s}^{-2}$

c $F = ma$ .... Notice that the force is in the opposite direction to motion so is negative.
$-500 = 100a$
$\Rightarrow a = -5 \, \text{m s}^{-2}$
$v^2 = u^2 + 2as$ .... Use equations of constant acceleration.
$0 = 20^2 + 2 \times (-5) \times s$
$s = 40 \, \text{m}$

You can see that the same force is exerted by both cyclists, but the acceleration is smaller for the larger mass. Mass can be said to be a measure of the amount of material present in an object, but can also be described as the reluctance of an object to change velocity.

EXERCISE 2A

1 Find the horizontal force required to make a car of mass 500 kg accelerate at $2 \, \text{m s}^{-2}$ on a horizontal road.

2 A wooden block of mass 0.3 kg is being pushed along a horizontal surface by a force of 1.2 N. Find its acceleration.

3 A gardener drags a roller on horizontal land with a force of 360 N, causing it to accelerate at $1.2 \, \text{m s}^{-2}$. Find the mass of the roller.

4 A man pushes a boy in a trolley, with total mass 60 kg, along horizontal land from rest with a constant force of 42 N for 10 s. Find the distance travelled in this time.

5 A snooker ball of mass 0.2 kg is struck so it starts moving at $1.2 \, \text{m s}^{-1}$. As it rolls, the table provides a constant resistance of 0.08 N. It strikes another ball 1 m away.

  a Find the speed with which it strikes the other ball.

  b What assumptions have you made when answering this question?

6 Find the constant force required to accelerate a mass of 5 kg from $3 \, \text{m s}^{-1}$ to $7 \, \text{m s}^{-1}$ in 8 s on a horizontal surface.

7 A ship of mass 20 tonnes is moving at $10\,\mathrm{m\,s^{-1}}$ when its engines stop and it decelerates in the water. It takes 500 m to come to rest. Find the resistance force, which is assumed to be constant, of the water on the ship.

8 A winch provides a constant force of 80 N and causes a block to accelerate on horizontal land from $2\,\mathrm{m\,s^{-1}}$ to $10\,\mathrm{m\,s^{-1}}$ in 6 s. Find the mass of the block.

9 A Formula 1 car and its driver have total mass of 800 kg. The driver is travelling at $100\,\mathrm{m\,s^{-1}}$ along a horizontal straight when, 100 m before a bend, he starts to brake, slowing to $40\,\mathrm{m\,s^{-1}}$. Assuming a constant braking force, find the force of the brakes on the car.

10 A strongman drags a stone ball from rest along a horizontal surface. He moves it 10 m in 4 s by exerting a constant force of 100 N. Find the mass of the stone ball.

11 A car and driver, of total mass 1350 kg, are moving at $30\,\mathrm{m\,s^{-1}}$ on a horizontal road when the driver sees roadworks 400 m ahead. She brakes, decelerating with a constant force of 600 N until arriving at the roadworks. Find the time elapsed before arriving at the roadworks.

12 A train travelling at $30\,\mathrm{m\,s^{-1}}$ on a horizontal track starts decelerating 360 m before coming to rest at a platform. The brakes provide a constant resistance of 100 kN. Find the mass of the train.

13 A block is being dragged along a horizontal surface by a constant horizontal force of size 45 N. It covers 8 m in the first 2 s and 8.5 m in the next 1 s. Find the mass of the block.

2.2 Combinations of forces

Newton's first and second laws refer to *net* or **resultant** forces because there may be more than one force acting on an object at any one time.

> **KEY POINT 2.3**
>
> Newton's second law is given more generally as
>
> Net force = mass × acceleration

TIP

It is normally easier to put all the information in a question into this equation rather than working out net force separately.

An object in **equilibrium** may have several forces acting on it, but their resultant is zero so it remains at rest or moving at constant velocity.

> **KEY POINT 2.4**
>
> Objects with a net force of zero acting on them are said to be in equilibrium and have no acceleration.

FAST FORWARD

You will consider problems with objects in equilibrium in Chapter 3.

You should use a diagram to work out which forces act in which directions. In this chapter the diagrams will be simple and include only one or two forces horizontally or vertically. You should get used to drawing diagrams for simple situations so that you are prepared for more complicated situations in later chapters.

Consider an aeroplane flying through the air. The forces acting on it are its weight, lift from the air on the wings, the driving force from the engine and propellers, and air resistance. Compare the following two illustrations.

TIP

Always draw a force diagram, even in simple situations, to ensure you have considered all forces in the problem and added them into the relevant equations.

Cambridge International AS & A Level Mathematics: Mechanics

Both illustrations have the same information, but the diagram on the right is simpler to draw and clearly shows the important information.

Diagrams are not pictures. In Mechanics you normally draw objects as circles or rectangles and show forces as arrows going out from the object. Objects being pushed from behind or dragged from in front are both shown as an arrow going forward from the object. The net force (or resultant force) is not shown on the diagram.

Accelerations are shown beside the diagram using a double arrow.

> **TIP**
>
> You do not usually include units on diagrams where forces are indicated by unknowns, otherwise there can be confusion about whether a letter refers to an unknown or a unit. Remember to use S.I. units throughout.

WORKED EXAMPLE 2.2

a A car of mass 600 kg has a driving force of 500 N and air resistance of 200 N. Find how long it takes to accelerate from $10\,\text{m}\,\text{s}^{-1}$ to $22\,\text{m}\,\text{s}^{-1}$.

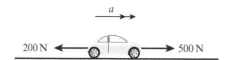

b The car stops providing a driving force and the brakes are applied. It decelerates from $22\,\text{m}\,\text{s}^{-1}$ to rest in 220 m. Find the force of the brakes.

Answer

a

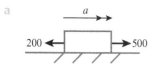

	The diagram is very simple and clear. The car is shown as a rectangle. The forces are arrows going out from the rectangle. The acceleration is a double arrow above the diagram. No resultant force is marked on the diagram.
$F = ma$ $500 - 200 = 600a$ $a = 0.5\,\text{m}\,\text{s}^{-2}$	All the horizontal forces make up the net force. Forces are negative if they are in the opposite direction to the motion.
$v = u + at$ $22 = 10 + 0.5t$ $t = 24\,\text{s}$	Acceleration is constant so you can use the relevant equation of constant acceleration to finish the problem.

b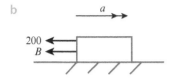

It is useful to draw new diagrams whenever the situation changes.
Note that the air resistance is still there, but the driving force is not there. The unknown B is now the braking force.
Note that acceleration is still in the direction of motion, although it is clear it will be negative because all the forces are acting the other way.

$v^2 = u^2 + 2as$
$0^2 = 22^2 + 2a \times 220$
$a = -1.1 \, \text{m s}^{-2}$

Use an equation of constant acceleration first to find the acceleration.
Three of the variables are known.

$F = ma$
$-200 - B = 600 \times (-1.1)$
$B = 460 \, \text{N}$

Use $F = ma$ now that two of the variables in this equation are known.

EXERCISE 2B

1 A boat and the sailor on it have a combined mass of 300 kg. The boat's engine provides a constant driving force of 100 N. It maintains a constant speed despite water resistance of 60 N and air resistance. Find the magnitude of the air resistance.

2 A boy and his friends have a tug-of-war with his father. His father pulls on one end of the rope with a force of 200 N. The boy and three friends each pull with equal force on the other end of the rope. The rope is in equilibrium. Find the force each child exerts on the rope.

3 A team of sailors pulls a boat over the sand to the sea. Each sailor is capable of providing a force of up to 300 N. The resistance from the sand is 2200 N. Find the minimum number of sailors needed on the team to maintain a constant speed.

4 A cyclist and her bike have a combined mass of 80 kg. She exerts a driving force of 200 N and experiences air resistance of 150 N. Find the acceleration of the cyclist.

5 A car of mass 1500 kg experiences air resistance of 450 N. It accelerates at $3 \, \text{m s}^{-2}$ on horizontal ground. Find the driving force exerted by the engine.

6 A boat of mass 2 tonnes has a driving force of 1000 N and accelerates at $0.2 \, \text{m s}^{-2}$. Find the resistance that the water provides.

7 A water-skier of mass 75 kg is towed by a horizontal rope with constant tension of 150 N. There is constant resistance from the water of 120 N. Find the time taken to reach a speed of $10 \, \text{m s}^{-1}$ from rest.

8 A learner driver is being tested on the emergency stop. The car has mass 1250 kg and is moving at $21 \, \text{m s}^{-1}$. When the driver presses the brakes there is a braking force of 10 000 N in addition to the air resistance of 500 N. Find the distance covered in coming to rest.

9 A small aircraft is accelerating on a runway. Its engines provide a constant driving force of 20 kN. There is average air resistance of 1000 N. The aircraft starts from rest and leaves the runway, which is 900 m long, with speed 80 m s^{-1}.

   a Find the mass of the aircraft.

   b What assumptions have been made to answer the question?

10 A drag racing car races on a track of length 400 m. The car has mass 600 kg and accelerates from rest with a constant driving force of 21 kN. There is air resistance of 1800 N. Find the speed of the car at the finish line of the race.

11 A wooden block of mass 6 kg is being dragged along a horizontal surface by a force of 10 N. It accelerates from 1 m s^{-1} to 3 m s^{-1} in 4 s. Find the size of the friction force acting on the wooden block.

12 A motorcyclist is travelling at 20 m s^{-1} on level road when she approaches roadworks and slows down to 10 m s^{-1} over a distance of 250 m. The combined mass of the motorcyclist and the motorcycle is 360 kg. There is air resistance of 80 N. Find the braking force of the motorcycle.

13 An aeroplane of mass 8 tonnes is flying horizontally through the air at 240 m s^{-1}. There is air resistance of 20 kN. The pilot reduces the driving force from the engines to slow to 160 m s^{-1} over 40 s before starting to descend. Find the magnitude of the reduced driving force.

14 A car of mass 1400 kg slows down from 30 m s^{-1} to 20 m s^{-1} when the driver sees a sign for reduced speed limit 400 m ahead. There is air resistance of 1000 N. Determine whether the driver needs to provide a braking force or just reduce the amount of driving force exerted, and find the size of the force.

2.3 Weight and motion due to gravity

If an object falls under gravity it moves with constant acceleration, whatever the mass of the object. This may seem contradictory because an object like a feather will fall to Earth much more slowly than a hammer. However, this is actually because of air resistance. Commander David Scott on Apollo 15 demonstrated that on the Moon, where there is no atmosphere, the two objects do land at the same time.

DID YOU KNOW?

Galileo Galilei (1564–1642) was the first to demonstrate that the mass does not affect the acceleration in free fall. It was thought he did this by dropping balls of the same material but of different masses from the Leaning Tower of Pisa to show they land at the same time. However, no account of this was made by Galileo and it is generally considered to have been a thought experiment. Actual experiments on inclined planes did verify Galileo's theory.

The acceleration in freefall due to gravity on Earth is denoted by the letter g and has a numerical value of approximately 10 m s^{-2}.

If an object of mass m kg falls under gravity with acceleration g m s^{-2}, then the force on the object due to gravity must be $F = mg$. This force is called the weight of an object and always acts towards the centre of the Earth, or vertically downwards in diagrams.

Chapter 2: Force and motion in one dimension

 KEY POINT 2.5

The weight of an object of mass m kg is given by $W = mg$. It is the force due to gravity, so is measured in newtons.

MODELLING ASSUMPTIONS

The value of g is actually closer to 9.8 m s^{-2}, but even that varies slightly depending on other factors. Because of the rotation of the Earth, the acceleration of an object in freefall is lower at the equator than at the poles. Gravity is also weaker at high altitudes and may even be weaker at depths inside the Earth. There can also be very slight local variations; for example, due to being near large mountains of dense rock. For the purposes of this course, we will assume that g is 10 m s^{-2}.

Above the surface of the Earth, the force due to gravity decreases. The difference is negligible for small distances, but this becomes important in space. In deep space, the gravitational pulls of not only the Earth but also the Sun become negligible. Under Newton's first law, objects like Voyager 1 and Voyager 2 in the far reaches of the Solar System will continue to move with the same velocity until they reach close enough to another star to feel its gravitational effect.

WORKED EXAMPLE 2.3

a A ball of mass 0.2 kg is thrown vertically upwards out of a window 4 m above the ground. The ball is released with speed 8 m s^{-1}. Assuming there is no air resistance, find how long it takes to hit the ground.

b If instead there is a constant air resistance of 0.1 N against the direction of motion, find how long the ball takes to hit the ground.

Answer

a Taking upwards as positive: — Define clearly which direction is positive.

$a = -g$

0.2g N

With only gravity acting, the acceleration is $-g$ if upwards is positive.

$s = ut + \frac{1}{2}at^2$

$-4 = 8t + \frac{1}{2} \times (-10)t^2$

$5t^2 - 8t - 4 = 0$

$t = 2 \text{ or } -0.4$

t is positive so the time to hit the ground is 2 s.

On the way up and on the way down, there is no change in the forces, so the whole motion can be dealt with as a single motion with acceleration $-g$.

b On the way up:

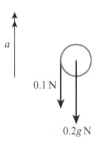

$$F = ma$$
$$-0.2g - 0.1 = 0.2a$$
$$a = -10.5$$

$$v = u + at$$
$$0 = 8 + (-10.5)t_1$$
$$t_1 = 0.762$$

So the time to reach highest point is 0.762 s (to 3 significant figures).

$$v^2 = u^2 + 2as$$
$$0^2 = 8^2 + 2 \times (-10.5)s_1$$
$$s_1 = 3.05$$

So distance travelled upwards is 3.05 m (to 3 significant figures).

On the way down:

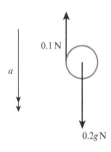

$$F = ma$$
$$0.2g - 0.1 = 0.2a$$
$$a = 9.5$$

$$s = ut + \frac{1}{2}at^2$$
$$s_1 + 4 = 0t_2 + \frac{1}{2} \times 9.5 t_2^2$$

giving $t_2 = 1.22$ (to 3 significant figures)

$t = t_1 + t_2 = 1.98$ s (3 significant figures)

> Separate the way up from the way down because resistance forces oppose motion, so the forces are different on the way down.
>
> You first need to find how long it takes to reach the highest point and the distance travelled in this time.

> Use Newton's second law to find the acceleration.

> The maximum height is reached when the velocity is reduced to zero.

> You will need to know the height reached so you can think about the motion on the way down.
>
> It is better to use given values rather than calculated values as much as possible.

> Draw a new diagram for the new situation.
>
> The motion is now downwards and the resistance force is in the opposite direction to the motion, so it now acts upwards.

> The ball is moving down for this stage, so we can define downwards as positive.

> Use the equation of constant acceleration with the total distance, including the maximum height s_1 found previously.

> Note that answers to previous calculations are written to 3 significant figures but you should always use values from the calculator, using Answer key or memories, in later calculations to avoid premature rounding errors.

WORKED EXAMPLE 2.4

a A ball is dropped from a height of 30 m above the ground. Two seconds later, another ball is thrown upwards from the ground with a speed of $5\,\mathrm{m\,s^{-1}}$. They collide at a time t s after the first ball was dropped. Find t.

b The balls collide at a height h m above the ground. Find h.

Answer

a Taking upwards as positive:

The mass is unknown but labelled as m_1, although this does not affect the acceleration because only gravity acts on the ball.

Let t_1 be the time after the first ball is dropped and s_1 be its displacement from its starting position.

$s = ut + \frac{1}{2}at^2$

$s_1 = 0t_1 + \frac{1}{2} \times (-10)t_1^2$

$h = 30 - 5t^2$

Measure height from the ground and time from the time when the first ball is dropped.

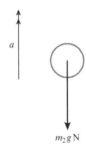

The second ball has a similar diagram, also with an unknown, but possibly different, mass.

Let t_2 be the time after the second ball is thrown and s_2 be its displacement from its starting position.

$s = ut + \frac{1}{2}at^2$

$s_2 = 5t_2 + \frac{1}{2} \times (-10)t_2^2$

$h = 5(t-2) - 5(t-2)^2$

Again, measure height from the ground and time from the time when the first ball is dropped.

$5(t-2) - 5(t-2)^2 = 30 - 5t^2$

$25t = 60$

$t = 2.4\,\mathrm{s}$

Solve the equations simultaneously.

b $h = 30 - 5t^2 = 30 - 5 \times 2.4^2 = 1.2\,\mathrm{m}$

Substitute into the equation for height.

EXERCISE 2C

1. Find the weight of a man of mass 70 kg.

2. A young goat has weight 180 N. Find its mass.

3. A ball is dropped from a height of 20 m. Find the time taken to hit the ground.

4. A water fountain projects water vertically upwards with initial speed $10\,\text{m s}^{-1}$. Find the maximum height the water reaches.

5. A ball is thrown vertically downwards with speed $5\,\text{m s}^{-1}$ from a height of 10 m. Find the speed with which it hits the ground.

6. A wrecking ball of mass 1 tonne is dropped onto a concrete surface to crack it. It needs to strike the ground at $5\,\text{m s}^{-1}$ to cause a crack. Find the minimum height from which it must be dropped.

7. A coin of mass 0.05 kg is dropped from the top of the Eiffel Tower, which is 300 m tall. It experiences air resistance of 0.01 N. Find the speed with which the coin hits the ground.

8. A winch lifts a bag of sand of mass 12 kg from the ground with a constant force of 240 N until it reaches a speed of $10\,\text{m s}^{-1}$. Then the winch provides a force to keep the bag moving at constant speed. Find the time taken to reach a height of 40 m.

9. A firework of mass 0.4 kg is fired vertically upwards with initial speed $40\,\text{m s}^{-1}$. The firework itself provides a force of 2 N upwards. The firework explodes after 6 s. Find the height at which it explodes.

10. A flare of mass 0.5 kg is fired vertically upwards with speed $30\,\text{m s}^{-1}$. The flare itself provides a force of 0.8 N upwards, even when the flare is falling, to keep the flare high for as long as possible. The flare is visible over the horizon when it reaches a height of 25 m.

    a Find how long the flare is visible for.

    b What assumptions have you made in your answer?

11. A feather of mass 10 g falls from rest from a height of 3 m and takes 2 s to hit the ground. Find the air resistance on the feather.

12. A ball of mass 0.3 kg is thrown upwards with speed $10\,\text{m s}^{-1}$. It experiences air resistance of 0.15 N. The ball lands on the ground 1.2 m below. Find the speed with which it hits the ground.

13. A bouncy ball is dropped from a height of 5 m. When it bounces its speed immediately after impact is 80% of the speed immediately before impact.

    a Find the maximum height of the ball after bouncing.

    b Show that the height is independent of the value used for g.

14. A parachutist of mass 70 kg falls out of an aeroplane from a height of 2000 m and falls under gravity until 600 m from the ground when he opens his parachute. The parachute provides a resistance of 2330 N. Find the speed at which the parachutist is travelling when he reaches the ground.

15 A ball is thrown vertically up at $10\,\mathrm{m\,s^{-1}}$. One second later another ball is thrown vertically up from the same point at $8\,\mathrm{m\,s^{-1}}$. Find the height at which the balls collide.

16 A pebble is dropped from rest into a deep well. At time $t\,\mathrm{s}$ later it splashes into the water at the bottom of the well. Sound travels at $340\,\mathrm{m\,s^{-1}}$ and is heard at the top of the well $5\,\mathrm{s}$ after the pebble was released. Find the depth of the well.

17 A ball of mass $2\,\mathrm{kg}$ is projected up in the air from ground level with speed $20\,\mathrm{m\,s^{-1}}$. It experiences constant air resistance R. It returns to ground level with speed $15\,\mathrm{m\,s^{-1}}$. Find R.

2.4 Normal contact force and motion in a vertical line

When an object rests on a table, why does it not fall? There is a force due to gravity, so there must be another force in the opposite direction keeping it in equilibrium. This is called the **reaction force**.

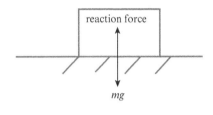

The reaction force is the force on an object from the surface it is resting on. It is usually denoted by the letter R. It is perpendicular to the surface it is in contact with, so sometimes N is used for the **normal contact force**.

KEY POINT 2.6

The normal contact force between the object and the surface it is on is called the reaction force and is always perpendicular to the surface.

When the object is on a horizontal surface, the normal contact force is usually the same magnitude as the weight. It simply prevents the object leaving or falling through the surface. However, when the surface is tilted with the object on it, or when other forces act on the object pushing it into the surface or pulling it away from the surface, the normal contact force is not usually the same magnitude as the weight.

FAST FORWARD

Some people mistakenly think the normal contact force is equal and opposite to the force of gravity. You will look at Newton's third law in Chapter 5, which is about forces that are equal and opposite. There *is* a force that is equal and opposite to the force of gravity on an object, but it acts on the Earth, not the object. Because the mass of the Earth is so large, the acceleration caused by the force is usually negligible. However, when you consider the motion of planets, the effect of gravity on both the Earth and other planets is important.

> **DID YOU KNOW?**
>
>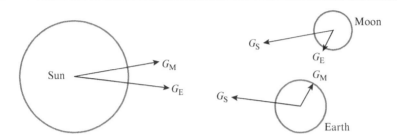
>
> In 1887, King Oscar II of Sweden and Norway established a prize for anyone who could solve the three-body problem, which asks what happens to a system of three objects each with gravity acting on them from each other, like the Sun, the Earth and the Moon, as shown in the diagram. Henri Poincaré (1854–1912) showed that, although we know the equations for the objects, there is no way to solve them. Moreover, he showed that if there is the slightest change in the initial positions or velocities of the bodies, the outcome may be entirely different. This led to the development of chaos theory and this effect became known as the butterfly effect. Another example of this is the weather, which is why it's so difficult to predict.

MODELLING ASSUMPTIONS

> If an object is on a table you may expect the table top to bend or even break if the object is heavy enough. You will assume that this is never the case and that the forces will never cause the surface to bend or break.
>
> As an object is lifted off the surface the normal contact force is reduced. When the force is reduced to zero, you would expect the object to lose contact with the surface and be lifted off. However, there are some cases where this does not happen in the real world. Vacuum suction pads, for example, can provide a force pulling the object towards the surface, as can electrostatic forces or sticky surfaces. You will ignore these possibilities in this course.
>
> Therefore, you will assume a normal contact force will be non-negative and there is no limit to how large it can be.

WORKED EXAMPLE 2.5

a A crane is lifting a pallet on that rests a stone block of mass 5 kg. The motion is vertically upwards. The crane lifts the pallet from rest to a speed of $3\,\text{m s}^{-1}$ in 6 m. Find the normal contact force on the stone block during the acceleration.

b If the normal contact force exceeds 650 N, the pallet may break and so this situation is considered unsafe. Assuming the same acceleration as in part **a**, find how many stone blocks the crane can lift safely.

Answer

a   Taking upwards as positive:

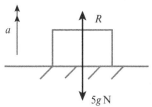

$v^2 = u^2 + 2as$ ..... Use an equation of constant acceleration to find the acceleration.

$3^2 = 0^2 + 2a \times 6$

$a = 0.75 \, \text{m s}^{-2}$

$F = ma$ ..... Use Newton's second law to find the contact force.

$R - 5g = 5 \times 0.75$

$R = 53.75 \, \text{N}$

The contact force is labelled R, but there are no units.

b   Taking upwards as positive:

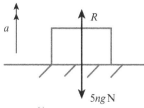

$F = ma$

$R - 5ng = 5n \times 0.75$

$R < 650$ ..... The restriction is given as an inequality.

$5ng + 5n \times 0.75 < 650$

$n < 12.1$

Hence, the maximum number of blocks is 12.

We define the number of stone blocks as n, but still consider the blocks as a single object.

Note that the question asked for a number of blocks, so it is the largest integer satisfying the inequality.

EXPLORE 2.1

An electronic scale and an object can be used to measure the acceleration in an elevator.

Use the scale to find the mass of the object. The scale works out the mass by measuring the contact force and dividing by g.

Put the object on the scale on the floor of the elevator. As you go up and down in the elevator, the reading on the scale should change.

As the elevator goes up, write down the maximum and minimum reading on the scale. The normal contact force can be calculated by multiplying by g. Use $F = ma$ for the object to now calculate the maximum acceleration and deceleration of the elevator.

Try this again when the elevator is going down.

EXERCISE 2D

1. An elevator is carrying a man of mass 70 kg upwards, accelerating at constant acceleration from rest to $10\,\text{m s}^{-1}$ in 2 s. Find the size of the normal contact force on the man.

2. An elevator is carrying a woman of mass 55 kg upwards. It is travelling upwards at $8\,\text{m s}^{-1}$ and starts to slow down at a constant rate when it is 9 m from where it stops. Find the size of the normal contact force on the woman.

3. An elevator is carrying a trolley of mass 30 kg downwards, accelerating at a constant rate from rest to $7\,\text{m s}^{-1}$ in 2 s. Find the size of the normal contact force on the trolley.

4. An elevator is carrying a child of mass 40 kg downwards. It is travelling at $8\,\text{m s}^{-1}$ and starts to slow down at constant acceleration when it is 6.25 m from where it stops. Find the size of the normal contact force on the child.

5. A forklift truck carries a wooden pallet. On the pallet is a box of tiles with mass 35 kg. The truck lifts the pallet and tiles with an initial acceleration of $2\,\text{m s}^{-2}$. Find the normal contact force on the tiles.

6. A weightlifter is trying to lift a bar with mass 200 kg from the floor. He lifts with a force of 1800 N but cannot lift it off the floor. Find the size of the normal contact force from the floor on the bar while the weightlifter is trying to lift the bar.

7. A plate of mass 0.7 kg is being held on a horizontal tray. The tray is lifted from rest on the floor and accelerates at a constant rate until it reaches a height of 1.25 m after 5 s. Find the normal contact force on the plate.

8. A man of mass 75 kg is standing in the basket of a hot-air balloon. The balloon is rising at $5\,\text{m s}^{-1}$ and 4 s later it is descending at $3\,\text{m s}^{-1}$. Assuming constant acceleration, find the normal contact force on the man.

9. A girl of mass 45 kg is sitting in a helicopter. The helicopter rises vertically with constant acceleration from rest to a speed of $40\,\text{m s}^{-1}$ when it reaches a height of 200 m.

   a Find the normal contact force on the girl.

   b What assumptions have been made to answer the question?

10. A drone carries a parcel of mass 4 kg. The parcel is held in place by two pads, on its top and on its bottom. The drone hovers at a height of 30 m before descending for 2 s to a height of 8 m. Find the normal contact force acting on the parcel as it descends and determine whether the force acts from the top pad or the bottom pad.

Checklist of learning and understanding

- A force is something that influences the motion of an object. Its size is measured in newtons (N).
- Force is related to acceleration by the equation: net force = mass × acceleration.
- Objects acted on only by the force of gravity have an acceleration of $g\,\text{m s}^{-2}$.
- The weight of an object is the force on it due to gravity and has magnitude $W = mg$.
- The reaction force or normal contact force is the force on an object due to being in contact with another object or surface. It acts perpendicular to the surface and is usually denoted by R (or sometimes N).

END-OF-CHAPTER REVIEW EXERCISE 2

1. A cyclist travelling on a horizontal road produces a constant horizontal force of 40 N. The total mass of the cyclist and the bicycle is 80 kg. Considering other forces to be negligible, find the distance covered as the cyclist increases her speed from $10\,\text{m\,s}^{-1}$ to $12\,\text{m\,s}^{-1}$.

2. A bag of sand of mass 20 kg is lifted on a pallet by a crane. The bag is lifted from rest to a height of 5 m in 8 s at constant acceleration. Find the normal contact force on the bag of sand.

3. A rower starts from rest and accelerates to $4\,\text{m\,s}^{-1}$ in 20 s. The combined mass of the rower and the boat is 100 kg. The rower provides a constant horizontal driving force of 60 N but is held back by a constant resistance from the water. Find the size of the resistance force.

4. A stone of mass 0.3 kg is dropped from the top of a cliff to the sea 40 m below. There is constant air resistance of 0.4 N as it falls.

   a Find the speed with which the stone hits the water.

   b The sound of the stone hitting the sea travels at $340\,\text{m\,s}^{-1}$. Find the time between releasing the stone and hearing the sound at the top of the cliff.

5. A train of mass 9000 kg is on a horizontal track. Its engine provides a constant driving force of 4000 N. There is constant air resistance of 400 N.

   a Find the time taken to reach a speed of $48\,\text{m\,s}^{-1}$ from rest.

   b When travelling at $48\,\text{m\,s}^{-1}$ the train enters a horizontal tunnel 400 m long. In the tunnel air resistance increases to 1000 N. Find the speed at which the train leaves the tunnel.

6. A submarine has mass 20 000 tonnes. With the engines on full power it can travel at $11\,\text{m\,s}^{-1}$ on the surface and $14\,\text{m\,s}^{-1}$ underwater.

   a When at maximum speed on the surface, the engines are turned off and it takes 4 km to come to a stop. Find the resistance from the water on the submarine.

   b Assuming the same resistance from the water, find the distance it would take to stop from maximum speed underwater when the engines are turned off.

   c Why is it not a reasonable assumption that the resistance underwater is the same as the resistance when the submarine is at the surface?

7. A diver of mass 60 kg dives from a height of 10 m into a swimming pool. Through the air there is resistance of 50 N.

   a Find the speed at which the diver enters the water.

   b Once in the water, the water provides an upwards force of 2000 N. Find the greatest depth in the water the diver reaches.

8. A car of mass 400 kg is approaching a junction and needs to stop in 40 m. It is travelling at $15\,\text{m\,s}^{-1}$ and there is air resistance of 1200 N. Determine whether the car needs to brake or accelerate and find the size of the relevant force.

9. A ball of mass 0.1 kg is projected vertically upwards from ground level with speed $9\,\text{m\,s}^{-1}$. It reaches a height of 3 m. There is air resistance against the motion.

   a Find the size of the air resistance.

   b Find the speed with which the ball hits the ground.

10 A car of mass 350 kg is travelling at 30 m s^{-1} when it starts to slow down, 100 m from a junction. At first, it slows just using the air resistance of 200 N. Then, at a distance of s m from the junction, it slows using brakes, providing a force of 2000 N as well as the air resistance. Find the distance from the junction at which the brakes must be applied if the car is to stop at the junction.

11 A firework of mass 0.3 kg has a charge that provides an upward force of 7 N for 3 s.

 a Assuming no air resistance, find the maximum height reached by the firework.

 b The explosive for the firework has a fuse that burns at a rate of 12 mm per second. Find how long the fuse should be so the firework explodes at the maximum height.

12 A boy drags a cart of mass 5 kg with force 10 N along a horizontal road. There is air resistance of 2 N. At some point the boy lets go of the cart and the cart slows down due to air resistance until coming to rest. In total, the cart has travelled 36 m. Find the length of time the boy was dragging the cart.

13 A light pallet is at rest on the ground with a stone of mass 30 kg on top of it but not attached. A crane lifts the pallet by providing a force of 310 N upwards to a height of 8 m, at which point the pallet instantly stops and the stone loses contact with it. Find the maximum height reached by the stone.

14 An air hockey table is 2 m long. A puck of mass 50 g is on the table at the middle point. A player hits the puck with initial speed 4 m s^{-1} directly towards one side. Once it is moving there is air resistance of R N. Every time the puck hits a side, the speed is reduced by 20%.

 a Show that if $R < \dfrac{32}{205}$, the puck returns past the middle point of the table.

 b Given that the puck does not return to the middle point a second time, find a lower bound for R.

15 A particle P is projected vertically upwards, from a point O, with a velocity of 8 m s^{-1}. The point A is the highest point reached by P. Find

 i the speed of P when it is at the mid-point of OA, [4]

 ii the time taken for P to reach the mid-point of OA while moving upwards. [2]

 Cambridge International AS & A Level Mathematics 9709 Paper 43 Q3 November 2012

16 Particles P and Q are projected vertically upwards, from different points on horizontal ground, with velocities of 20 m s^{-1} and 25 m s^{-1} respectively. Q is projected 0.4 s later than P. Find

 i the time for which P's height above the ground is greater than 15 m, [3]

 ii the velocities of P and Q at the instant when the particles are at the same height. [5]

 Cambridge International AS & A Level Mathematics 9709 Paper 42 Q5 November 2010

17 A particle of mass 3 kg falls from rest at a point 5 m above the surface of a liquid which is in a container. There is no instantaneous change in speed of the particle as it enters the liquid. The depth of the liquid in the container is 4 m. The downward acceleration of the particle while it is moving in the liquid is 5.5 m s^{-2}.

 i Find the resistance to motion of the particle while it is moving in the liquid. [2]

 ii Sketch the velocity–time graph for the motion of the particle, from the time it starts to move until the time it reaches the bottom of the container. Show on your sketch the velocity and the time when the particle enters the liquid, and when the particle reaches the bottom of the container. [7]

 Cambridge International AS & A Level Mathematics 9709 Paper 41 Q6 November 2014

Chapter 3
Forces in two dimensions

In this chapter you will learn how to:

- resolve forces in two dimensions
- find resultants of more than one force in two dimensions
- use $F = ma$ in two directions
- find directions of motion and accelerations.

Cambridge International AS & A Level Mathematics: Mechanics

> **PREREQUISITE KNOWLEDGE**
>
> | Where it comes from | What you should be able to do | Check your skills |
> |---|---|---|
> | IGCSE / O Level Mathematics | Use Pythagoras' theorem. | 1 Find the hypotenuse of a right-angled triangle with short sides of length 5 m and 12 m. |
> | IGCSE / O Level Mathematics | Use trigonometry for right-angled triangles. | 2 A triangle, ABC, has a right angle at B. Length AC is 8 m and $\angle BAC$ is $40°$. Find lengths AB and BC. |
> | IGCSE / O Level Mathematics | Use the sine rule and the cosine rule. | 3 A triangle, ABC, has length AC 6 m, $\angle BAC$ $40°$ and length BC 7 m. Find $\angle ABC$ and length AB. |
> | Pure Mathematics 1 | Use the trigonometry identity $\sin^2 \theta + \cos^2 \theta = 1$ | 4 If $\sin \theta = \dfrac{3}{5}$, find $\cos \theta$. |
> | Pure Mathematics 1 | Use the trigonometry identity $\dfrac{\sin \theta}{\cos \theta} = \tan \theta$ | 5 If $\sin \theta = \dfrac{5}{13}$, find $\tan \theta$. |

How do you combine forces that are not acting in the same line?

Imagine two children are playing with a toy. They both pull it with a force of 10 N. What would be the net force? Before you can answer this question, you need to know the directions in which the forces are acting. If both children want to take the toy to the same place and their forces act in the same direction, the net force would be 20 N. If they are trying to take the toy away from each other and their forces act in opposite directions, there would be no net force. But what if the forces are not parallel? For example, one could be to the north and one to the east.

This chapter covers how to solve problems with forces in two dimensions.

3.1 Resolving forces in horizontal and vertical directions in equilibrium problems

A force is a vector quantity. When vectors are added it is the equivalent of joining one vector on to the end of the other.

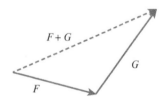

This property can be used in reverse by splitting a vector into the sum of two others called **components**. You choose the two vectors to be in perpendicular directions to make it possible to set up equations. The components and the original vector will then always form a right-angled triangle, so you can find the values of each component using trigonometry for right-angled triangles or Pythagoras' theorem.

You can use the trigonometric relationships $\sin \theta = \dfrac{\text{opposite}}{\text{hypotenuse}}$ and $\cos \theta = \dfrac{\text{adjacent}}{\text{hypotenuse}}$ to find how the components of a force relate to the original force. In the diagram:

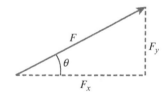

$$F_x = F \cos \theta$$

$$F_y = F \sin \theta$$

Note that if you knew the other angle in this triangle, you would have to use sin to find F_x and cos to find F_y.

Chapter 3: Forces in two dimensions

KEY POINT 3.1

The components of a force are at right angles to each other. The original force is the hypotenuse of the triangle.

TIP

When drawing force diagrams, you can draw the triangles to help work out the components, but it is best not to mark the components as separate forces or you may count the force twice.

Components are not extra forces. They are the parts of a force already given, which act in certain directions.

Equations are formed by finding the net component horizontally and the net component vertically. This is called **resolving** the forces in each direction.

DID YOU KNOW?

The shape that a chain, wire or rope makes when it hangs between two points has a mathematical formula. It is called the catenary curve after the Latin word for chain. You can resolve for each link in the chain, or particle on a wire or rope, to form differential equations. You can then solve them to get the equation of the curve. The formula for the curve is a hyperbolic function (derived from the exponential function), but a small part of the curve looks very similar to a parabolic curve, like those for quadratic graphs.

KEY POINT 3.2

In equilibrium the net force in both perpendicular directions will be zero.

WORKED EXAMPLE 3.1

A particle of mass 4 kg is held in place by a force of magnitude 100 N acting at an angle θ above the horizontal and a horizontal force of F N. Find the values of θ and F.

Answer

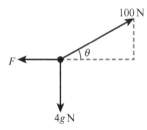

The dashed lines show the horizontal and vertical directions to allow calculation of the components of the 100 N force.

$100 \sin \theta = 4g$ Resolving vertically.

$\theta = 23.6°$

$F = 100 \cos \theta$ Resolving horizontally.

$= 91.7$ There are no units as this is the value of F.

Cambridge International AS & A Level Mathematics: Mechanics

WORKED EXAMPLE 3.2

A boat is held in place by a force of 5 N due east, a force of 10 N due south and a force F N, on a bearing of θ. Find the values of F and θ.

Answer

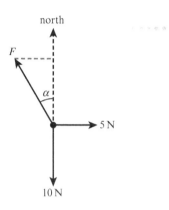

To be in equilibrium F must have a component to the west to cancel out the 5 N force and a component to the north to cancel out the 10 N force.

A triangle is drawn to make it easier to work out the components, but the components are not marked.

Bearings are always measured clockwise from north. If the bearing is not acute it is often easier to mark an acute angle, here α, relative to one of the four basic. Here the bearing $\theta = 360° - \alpha$.

$F \sin \alpha = 5$ — Resolving east-west.

$F \cos \alpha = 10$ — Resolving north-south.

$\tan \alpha = \dfrac{1}{2}$ — Dividing the equations.

$\alpha = 26.6°$

Therefore $\theta = 333.4°$.

$F^2 = 5^2 + 10^2$ — By Pythagoras' theorem.

$F = 11.2$

EXERCISE 3A

1 Find the components of the forces in the diagrams:

   a horizontally, specifying whether it is left or right

   b vertically, specifying whether it is upwards or downwards.

   i 15 N at 40° above horizontal

   ii 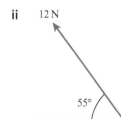 12 N at 55°

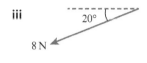

iii

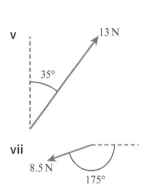

v

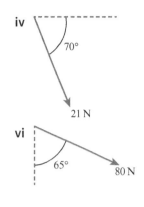

iv

vi

vii

viii 120 N

2 a A force, F, has a horizontal component, F_x, of 10 N and acts at 20° above the rightwards horizontal, as shown in the diagram. Find F and the vertical component, F_y.

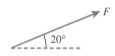

   b A force, F, has a vertical component, F_y, of 8 N and acts at 25° to the right of the upwards vertical. Find F and the horizontal component, F_x.

   c A force, F, has a vertical component, F_y, of 8 N and a horizontal component, F_x, of 10 N. Find F and the angle, θ, that the force makes with the rightwards horizontal.

   d A force of 25 N has a horizontal component, F_x, of 17 N and acts above the horizontal. Find the vertical component, F_y, and the angle, θ, above the rightwards horizontal at which the force acts.

   e A force of 3.8 N has a vertical component, F_y, of 3 N and acts to the left of the vertical. Find the horizontal component, F_x, and the angle, θ, above the leftwards horizontal at which the force acts.

3 A particle in equilibrium has three forces of magnitudes 5 N, 6 N and F N acting on it in the horizontal plane in the directions shown. Find the values of F and θ.

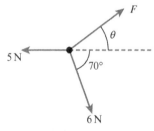

4 A lightshade of mass 2 kg is hung from the ceiling by two strings. One is fixed with tension 8 N at 20° to the vertical. The other is fixed with tension T N at an angle θ to the vertical.

   a By modelling the lightshade as a particle, draw a force diagram for this situation.

   b Resolve horizontally to find a value for $T \sin \theta$ and resolve vertically to find a value for $T \cos \theta$.

   c Hence, find the values of T and θ.

5 A ship is being blown by a breeze with a force of 100 N on a bearing of 280°, as shown in the diagram. It is pulled by a rope attached to the shore with force 50 N on a bearing of 170°. A tugboat holds it in place. Find the size and bearing of the force F applied by the tugboat.

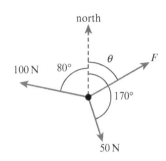

6 A wooden block of weight 20 N is at rest on a horizontal surface. It is pulled by a force of 30 N acting at 10° above the horizontal, as shown in the diagram, and remains at rest because of a horizontal frictional force, F.

   a Draw the force diagram for this situation.

   b Find the size of F and the size of the normal contact force.

7 A winch is dragging a caravan along a horizontal road at constant velocity. The caravan has mass 750 kg. The winch provides a force of 850 N and acts at angle θ above the horizontal, as shown in the diagram. There is friction of 700 N.

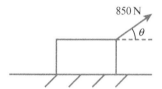

   a Draw the force diagram for this situation.

   b Find the value of θ and size of the normal contact force.

8 A box of weight 50 N is being dragged at constant velocity along a horizontal road by a force, F, acting at 15° above the horizontal. It experiences friction of 10 N.

   a Draw the force diagram for this situation.

   b Find F and the normal contact force.

9 A small aeroplane of mass 5000 kg is towed along a runway at constant speed by a rope acting at 20° below the horizontal. There is friction and air resistance horizontally with total force 4000 N. Find the tension in the rope and the normal contact force.

10 A wooden block is held in position by three horizontal forces, as shown in the diagram. One acts to the left with force 56 N. One acts with force F at an angle θ, where $\sin \theta = \dfrac{3}{5}$, above the rightwards horizontal. One acts with force G at an angle φ, where $\sin \varphi = \dfrac{15}{17}$, below the rightwards horizontal. Find F and G.

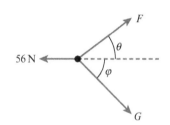

11 A block with weight 44 N is held in equilibrium by two ropes, one with tension, T_1, acting at angle $\sin^{-1} \dfrac{15}{17}$ to the upwards vertical and the other with tension, T_2, acting at angle $\sin^{-1} \dfrac{5}{13}$ to the upwards vertical. Find T_1 and T_2.

12 A box with weight 400 N is at rest on a horizontal surface. A man is pulling on a rope to try to get the box to move. The force he can exert depends on the angle at which he holds the rope, so that when the rope is at an angle θ above the horizontal, the force he exerts is $1600 \sin \theta$ N. He starts by holding the rope horizontally and gradually increases the angle, thereby increasing the force. Another man tries to prevent this motion of the box, by pulling horizontally. He can exert a maximum force of 700 N. Find the angle at which the box can no longer remain on the ground. Hence, determine whether the box lifts off the ground first or slides along the ground first.

13 A particle has three forces acting on it, as shown in the diagram, where $\sin\theta = \dfrac{3}{5}$.

Show that $F + G = 150\sqrt{3}$ by resolving horizontally, and write down another equation by resolving vertically. Hence, show that $G = 75\sqrt{3} + 100$ and find F.

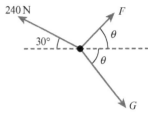

14 A particle has three horizontal forces acting on it, as shown in the diagram.

Show that $\cos\alpha = \dfrac{14 - 13\cos\beta}{15}$ and find an expression for $\sin\alpha$. Use $\cos^2\alpha + \sin^2\alpha = 1$ to get an equation in β. Hence, find α and β.

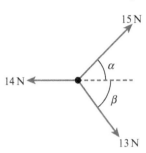

3.2 Resolving forces at other angles in equilibrium problems

Try resolving horizontally and vertically for the forces in equilibrium in this diagram.

You should get the two equations:

$R\cos 65° = T\cos 25°$

$R\sin 65° + T\sin 25° = 10$

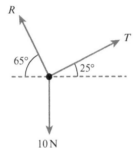

If there are two unknown forces and neither of them is vertical or horizontal, resolving horizontally and vertically will lead to two equations, both of which involve two unknowns.

You can solve these equations simultaneously, but it could be challenging. It would be easier if one equation involved only one unknown.

Sometimes it is easier to resolve forces in directions other than horizontal and vertical.

A force has no component in the direction perpendicular to its **line of action**. This means that if you resolve perpendicular to an unknown force, the unknown force will not appear in the equation.

If you resolve in a direction perpendicular to R in the example illustrated, R will not appear in the equation so you can solve directly for T. You will need to find the component of the 10 N force in this direction.

As an alternative to drawing a right-angled triangle, it may be easier to consider the angle between the force and the direction in which you are resolving. When resolving parallel to a certain direction, as marked by p in the following diagram, the component of the force F in that direction will be adjacent to the angle θ between the force and the direction p. Therefore, the component F_p is found using the cosine of the angle.

> **TIP**
>
> In problems that involve a slope, you should resolve forces parallel and perpendicular to the slope. In other cases, choose directions perpendicular to an unknown force. Choose the directions carefully so there are as few unknowns as possible in each direction, to make solving the equations easier.

KEY POINT 3.3

The component of a force, F, parallel to a given direction, p, can be found by $F_p = F \cos\theta$, where θ is the angle between the force and the direction p.

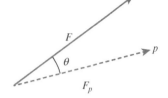

WORKED EXAMPLE 3.3

A boat is held in equilibrium by three forces of 10 N, F N and 20 N, as shown in the diagram. Find the values of F and θ.

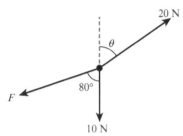

Resolving horizontally and vertically will leave awkward simultaneous equations in F and θ.

Since F is an unknown force, we resolve perpendicular to F so it does not appear in the equations, to find θ.

Then you can find F.

Answer

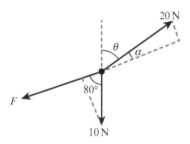

To help, dashed lines are added to the force diagram to create right-angled triangles, with the forces as the hypotenuses and the other two sides parallel and perpendicular to F.

Mark the angle α to compare the 20 N force with the direction of F.

You can find θ from α because they add up to 80°.

$20 \sin\alpha = 10 \sin 80°$

$\alpha = 29.5°$

Therefore $\theta = 80° - \alpha = 50.5°$

Resolve perpendicular to F.

Notice F does not appear in this equation.

$F + 10 \cos 80° = 20 \cos\alpha$

$F = 15.7$

Resolve parallel to F.

WORKED EXAMPLE 3.4

A block of mass 10 kg is held in equilibrium on a slope at an angle of 20° to the horizontal by a force, F, acting at 15° above the slope. Find F and the normal contact force.

Answer

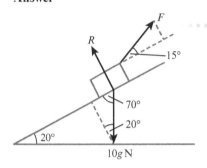

The normal contact (reaction) force is perpendicular to the slope.

If you draw the weight arrow down to the horizontal line from the bottom of the slope, it may make it easier to find missing angles.

You will be resolving perpendicular and parallel to the slope, so add dotted lines to form the right-angled triangles, making sure the forces are the hypotenuses of the triangles. This allows you to find components in the directions of the dotted lines.

$F \cos 15° = 10g \sin 20°$ — Resolve parallel to slope.
$F = 35.4 \text{ N}$ — Notice R does not appear in this equation.

$R + F \sin 15° = 10g \cos 20°$ — Resolve perpendicular to slope.
$R = 84.8 \text{ N}$

> **TIP**
>
> When you draw a diagram involving a slope, make sure the slope does not look like it is at 45°, as it will make it clearer if angles at other points in the diagram are the same as the angle of the slope or not.

> **WEB LINK**
>
> You may want to have a go at the *Make it equal* resource at the *Vector Geometry* station on the Underground Mathematics website.
>
> Note that you do not need to be able to use $\mathbf{i} - \mathbf{j}$ vector notation for this Mechanics syllabus.

EXERCISE 3B

1 Find the components of the following forces in the direction of the dashed arrow. Although it might seem clear from the diagram, make sure you specify whether the component is in the given direction or in the opposite direction.

   a 150 N, 35°

   b 14 N, 25°

   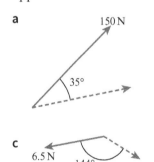

   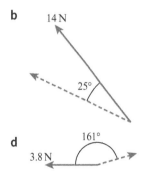

   c 6.5 N, 144°

   d 3.8 N, 161°

2 Find the components of the following forces perpendicular to the direction of the dashed arrow. Make it clear whether the component is in the perpendicular direction clockwise or anticlockwise from the direction given.

a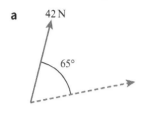

b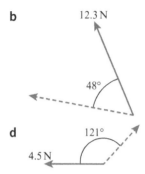

c 0.4 N
148°

d 4.5 N
121°

3 A particle has three forces acting on it, as shown in the diagram. By resolving perpendicular to and parallel to F, find F and θ.

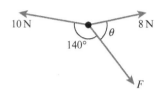

4 A boat is held in equilibrium by two tugboats. One pulls with a force of 100 N on a bearing of 190°. One pulls on a bearing of 340° with tension T. The wind blows with a force on the boat of F on a bearing of 50°. By resolving perpendicular to T, find F. Find also T.

5 A book of mass 3 kg is prevented from sliding down a slope at 15° to the horizontal by friction acting up the slope and parallel to it. Find the force of friction and the normal contact force.

6 A wooden block of mass 4 kg is held at rest on a slope at angle θ to the horizontal by a force of 12 N acting up the slope and parallel to it. Find the slope's angle and the normal contact force.

7 A particle of mass 2 kg is held in equilibrium on a slope at 13° to the horizontal by a force F acting at 10° to the slope above it. Find F and the normal contact force.

8 A box of mass 12 kg is held in equilibrium on a slope at 18° to the horizontal by a force of size 50 N acting at an angle θ above the slope. Find θ and the normal contact force.

9 A boy is dragging a bag of mass 8 kg up a slope at an angle of 17° to the horizontal and exerts a force of 50 N parallel to the slope to do this. Air resistance, F, parallel to the slope prevents the boy from increasing his speed, so he maintains a constant speed. Find the magnitude of the air resistance and the normal contact force.

10 A girl is dragging a sled of mass 20 kg up a slope at angle 14° to the horizontal. She pulls at an angle of θ above the slope with a force of 70 N. She maintains a constant speed despite friction of 10 N parallel to the slope. Find θ and the normal contact force.

11 A particle of mass 4 kg is at rest on a slope at an angle of 49° to the horizontal. There is a frictional force of 10 N acting up the slope and a force F going up the slope acting at 9° above the slope. Find F and the normal contact force.

PS 12 A heavy box of mass 50 kg is on a slope at angle 25° to the horizontal. There is no friction to prevent it sliding down the slope, but there are three rods attached, at 40°, 50° and 60° above the slope, for people to drag it. A man and two boys hold the rods to keep the box in equilibrium.

  a Show that, if the man pulls with a force of 170 N and each boy can pull with a force of up to 90 N, they can hold the box in equilibrium.

  b If instead the man pulls with force 180 N and each boy can pull with a force up to 70 N, determine whether or not they can hold the box in equilibrium and state which rod each should hold.

13 A box of mass 20 kg is on a horizontal surface. There are three rods attached on one side, at 10°, 25° and 35° above the horizontal, for people to drag it. Three people are available to pull on these rods and they are capable of providing forces of 150 N, 200 N and 250 N.

  a The box is being pulled in the opposite direction by a horizontal force of 425 N. Show that only two of the people are required to keep the box in equilibrium. State which of the rods each person holds.

  b The horizontal force is increased to 550 N. Show that if the box is to be prevented from moving horizontally, it cannot remain on the ground.

3.3 The triangle of forces and Lami's theorem for three-force equilibrium problems

The methods in this section are not required by the syllabus. However, they provide neat and efficient methods for solving some problems. Although the questions can all be solved using the methods from the previous sections, they may be solved more quickly using alternative methods involving the triangle of forces or Lami's theorem.

If three forces act on an object to keep it in equilibrium, they will have no resultant. This means that we can draw them end to end and they will finish where they started and form a triangle. We can then use trigonometry to solve the problem.

> **KEY POINT 3.4**
>
> By drawing a triangle of forces, we can use the sine rule or cosine rule directly without resolving components. The lengths of the sides will be the magnitudes of the forces.

First, draw the force diagram as a triangle of forces.

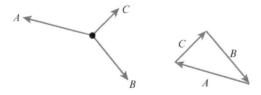

> **TIP**
>
> The forces can be drawn in a triangle of forces in any order. Choose an order where it is easiest to work out the angles.

You can add angles to the diagrams. You should extend the straight lines in the triangle, as shown in the diagram.

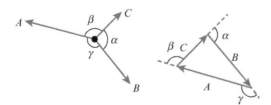

Applying the sine rule to the triangle gives $\dfrac{A}{\sin(180° - \alpha)} = \dfrac{B}{\sin(180° - \beta)} = \dfrac{C}{\sin(180° - \gamma)}$.

Since $\sin\theta = \sin(180° - \theta)$, this leads to $\dfrac{A}{\sin\alpha} = \dfrac{B}{\sin\beta} = \dfrac{C}{\sin\gamma}$.

KEY POINT 3.5

Lami's theorem states that for a particle in equilibrium with three forces on it, the ratio of the magnitude of the force with the sine of the angle between the other two forces is the same for each force.

$$\frac{A}{\sin \alpha} = \frac{B}{\sin \beta} = \frac{C}{\sin \gamma}$$

EXPLORE 3.1

Forces of size 5 N, 6 N and 12 N act on an object. Can the object be in equilibrium? Here are the opinions of two students.

Student A	Student B
There is no way of making two of them equal to the third, so they cannot cancel out, and the object cannot be in equilibrium.	If the forces were at different angles, it might be possible for it to be in equilibrium.

Is one of the students correct? If the forces were of different sizes, in which circumstances would each student be correct?

WORKED EXAMPLE 3.5

An object is in equilibrium by the action of forces of 10 N, 8 N and 9 N, as shown in the diagram. Find the values of θ and φ.

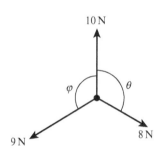

Answer

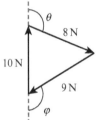

Redraw the diagram as a triangle of forces.

$\cos(180° - \varphi) = \dfrac{8^2 + 10^2 - 9^2}{2 \times 8 \times 10}$

$\varphi = 121.2°$

Use the cosine rule to find the angle between the 8 N and 10 N forces.

$\cos(180° - \varphi) = \dfrac{9^2 + 10^2 - 8^2}{2 \times 9 \times 10}$

$\varphi = 130.5°$

Use the cosine rule to find the angle between the 10 N and 9 N forces, and use the alternate angles theorem using the parallel 'north' lines.

WORKED EXAMPLE 3.6

A ship is held in equilibrium by ropes on bearings of 120° and 220°. The wind is blowing due north and exerting a force of 90 N on the ship. Find the tensions in the two ropes.

Answer

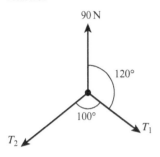

$$\frac{90}{\sin 100°} = \frac{T_1}{\sin 140°}$$

$T_1 = 58.7$ N .... Use Lami's theorem.

$$\frac{90}{\sin 100°} = \frac{T_2}{\sin 120°}$$

$T_2 = 79.1$ N .... Use Lami's theorem again.

WORKED EXAMPLE 3.7

A particle is held in equilibrium by three forces, as shown in the diagram. Find the sizes of F and α.

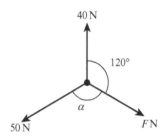

Answer

Using the method of resolving forces: .... Resolving perpendicular to F followed by resolving parallel to F.

$50\sin(180° - \alpha) = 40\sin 60°$

$\alpha = 43.9°$ or $136.1°$ .... Note that α must be bigger than 60° or there would be no component of the forces to the left of the 40 N force so the particle could not be in equilibrium.

Here $\alpha = 136.1°$.

$F = 50\cos(180° - \alpha) + 40\cos 60° = 56.1$

$$\frac{40}{\sin \alpha} = \frac{50}{\sin 120°}$$ .... Using Lami's theorem.

$\alpha = 43.9°$ or $136.1°$

Here $\alpha = 136.1°$.

$$\frac{F}{\sin(360° - 120° - \alpha)} = \frac{50}{\sin 120°}$$

$\Rightarrow F = 56.1$

Using the method of the triangle of forces:

$$\frac{40}{\sin(180°-\alpha)} = \frac{50}{\sin(180°-120°)}$$ .... Using the sine rule.

$\alpha = 43.9°$ or $136.1°$
Here $\alpha = 136.1°$.
$\beta = 76.1°$
$F^2 = 40^2 + 50^2 - 2 \times 40 \times 50 \times \cos\beta$ .... Using the cosine rule.
$F = 56.1$

EXERCISE 3C

1. A particle is held in place by forces of 8 N, 11 N and 12 N, as shown in the diagram. Find the values of θ and φ.

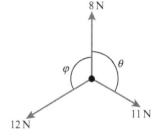

2. A mass of 5 kg is held in equilibrium by two ropes with tensions of 30 N and 40 N. Find the angles that the ropes make with the vertical.

3. A mass of 7 kg is held in equilibrium by two ropes. One has tension 20 N and acts at 40° to the upwards vertical. Find the tension in the other rope and the angle that it makes with the upwards vertical.

4. A ship is held in place by two ropes with forces 40 N and 35 N, as shown in the diagram, which prevent the wind blowing it away. The wind has force F and acts at an angle θ to the 35 N force, as shown. Find the sizes of θ and F.

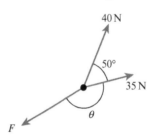

5. Three ropes pull a boat, which remains in equilibrium. The ropes act due north and on bearings of 100° and 210°. The one acting north has tension 25 N. Find the tensions in the other ropes.

6. A box has two ropes holding it in place. It is pushed by a force of 10 N. The angles between the force and the ropes are 120° and 150°. Find the tensions in the ropes.

7. An 8 N force, a 9 N force and a 10 N force on an object result in no net force. Find the angle between the 8 N and the 9 N forces.

8 A land yacht is a vehicle with a sail that gets blown by the wind, but it moves on solid ground. An adult and a child are holding ropes attached to the land yacht. The adult is capable of pulling with a force of 300 N. The child is capable of pulling with a force of 80 N. They cannot pull in the same direction or they get in each other's way, so there needs to be an angle of at least 20° between their ropes.

   a For what strength of wind can the two of them work together to prevent the yacht from moving?

   b For what strength of wind can the adult prevent the child from moving the boat?

   c When the wind is blowing with a force of 130 N, the adult pulls the land yacht directly against the wind. The child can cause the path of the yacht to deviate from the direction in which the adult pulls. Find the maximum angle of deviation the child can cause.

9 A particle is held in equilibrium by three forces. Two of the forces have the same size, F N. Prove that the third force acts along the line of the angle bisector of the lines of action of the other two forces.

10 Four forces on an object, A, B, C and D, result in no net force. If the angle between forces A and B is α and the angle between forces C and D is γ, show that $A^2 + B^2 + 2AB\cos\alpha = C^2 + D^2 + 2CD\cos\gamma$.

3.4 Non-equilibrium problems for objects on slopes and known directions of acceleration

When forces are not in equilibrium, the net force will not be zero, so we can apply Newton's second law. The object will accelerate.

REWIND

You used Newton's second law in Chapter 2, Section 2.1.

We calculate the acceleration using $F = ma$, but we need to resolve the forces into components in a relevant direction and find the net force in that direction.

We need to choose carefully which directions to resolve in. When an object is on a slope it is clear the object is not going to fly off the slope or go into the slope, so any acceleration will be parallel to the slope, either up it or down it. In this situation there will be no net force in the direction perpendicular to the slope, so we should resolve in directions perpendicular and parallel to the slope.

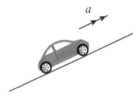

Alternatively, if a ship is being towed in a straight line by two tugboats, you may be able to see the direction of motion from the bearing of the ship. There will be no acceleration perpendicular to the direction of motion, so we should resolve in directions perpendicular and parallel to the motion.

In some situations, as well as the acceleration being unknown, one of the forces or an angle is also unknown. For example, suppose two people are pulling a car with ropes at known angles. The force from person A is known, but person B is pulling with enough force to keep the car following the path marked by the dotted line. Without knowing the size of the force, it is impossible to work out the acceleration from one equation.

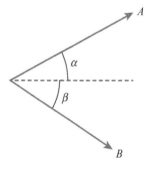

In this case, we must resolve forces in the perpendicular direction to get a second equation. We know that there is no acceleration in this direction, so this equation is set up in the same way as with equilibrium problems.

KEY POINT 3.6

The net force in the direction perpendicular to the acceleration is zero.

MODELLING ASSUMPTIONS

The scenarios in these questions involve net forces that cause acceleration. Did the forces instantly appear at those sizes? Forces like gravity will always be there, but someone pulling on a rope may have to increase the force from zero.

If that happens, why was there not a smaller acceleration while the force was increasing to the size given? There are different assumptions that may have been made to model the situation more easily, without significantly affecting the values calculated.

In some cases, the object is said to be held in place. That means there is initially some other force keeping the object in equilibrium. That force is instantaneously removed so the forces under consideration are already at the values given. In other cases, the time taken to reach the given force values is considered negligible, and it is modelled as if the forces are instantly at the values given.

We have also noted earlier that we are ignoring the shape of objects and considering them all to be particles. In many cases this does not have an impact because the object slides along a surface like a particle does. However, round objects like balls, wheels or cylinders can roll. This has an impact on the motion, but at this stage we will treat them as if they are particles, just sliding.

WORKED EXAMPLE 3.8

A box of mass 25 kg is dragged along the floor by a force of 30 N acting at 20° above the horizontal. Find the acceleration and the normal contact force.

Answer

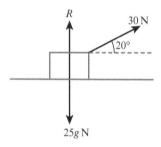

$F = ma$

$30 \cos 20° = 25a$ — Resolving horizontally.

$a = 1.13 \, \text{m s}^{-2}$

$R + 30 \sin 20° = 25g$ — Resolving vertically, there must be no resultant force, otherwise the box would leave the floor or sink into the floor.

$R = 240 \, \text{N}$

Chapter 3: Forces in two dimensions

WORKED EXAMPLE 3.9

A boat of mass 40 kg has an engine providing a driving force of 30 N in an easterly direction. It is also being blown by the wind with a force T to the north. The boat moves on a bearing of 60°. Find T and the acceleration of the boat.

Answer

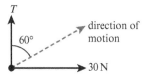

$T \sin 60° = 30 \sin 30°$

$T = 17.3\,\text{N}$

Resolve perpendicular to the direction of motion first because there will be no net force in this direction.

$F = ma$

$T \cos 60° + 30 \cos 30° = 40a$

$a = 0.866\,\text{m s}^{-2}$

Resolve in the direction of motion.

WORKED EXAMPLE 3.10

A table is sliding down a slope at an angle 20° to the horizontal. There is resistance of 10 N acting up the slope parallel to it. The table takes 5 s to slide 10 m down the slope from rest. Find the mass of the table.

Answer

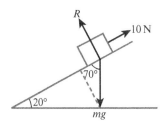

The table is modelled as a particle, so we do not worry about its shape for the diagram.

$s = ut + \dfrac{1}{2}at^2$

$10 = \dfrac{1}{2} a \times 5^2$

$a = 0.8\,\text{m s}^{-2}$

Use information given to find the acceleration first.

$F = ma$

$mg \sin 20° - 10 = m \times 0.8$

$m = 3.82\,\text{kg}$

Resolve parallel to the slope.

Here F is the net force and, since we are taking the direction of motion as positive, the 10 N force is negative.

> **REWIND**
>
> Look back to Chapter 1, Section 1.3, if you need a reminder of the equations of constant acceleration.

> **WEB LINK**
>
> You may want to have a go at the *Make it stop* resource at the *Vector Geometry* station on the Underground Mathematics website.

EXERCISE 3D

1. A wooden block of mass 5 kg is on a horizontal surface. It is dragged by a force of 20 N acting at 14° above the horizontal, as shown in the following diagram. Find the acceleration of the block and the normal contact force.

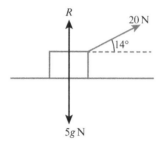

2. A book of mass 2 kg is dragged along a horizontal surface by a rope at 6° above the horizontal. It accelerates at $0.3\,\mathrm{m\,s^{-2}}$.

   a Draw the force diagram for this situation.

   b Find the tension in the rope and the normal contact force.

3. A box of mass 10 kg is pulled along a horizontal surface by a rope with tension 20 N at an angle θ above the horizontal. The box accelerates at $1.2\,\mathrm{m\,s^{-2}}$. Find θ and the normal contact force.

4. A car of mass 1000 kg is being towed by two people holding ropes. One pulls with a tension of 80 N at an angle of 18° to the direction of motion. The other pulls at an angle of 25° to the direction of motion, as shown in the diagram. Find the tension, T, in the second rope and the acceleration of the car.

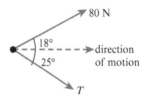

5. A box of mass 20 kg is dragged by a force of 40 N at an angle of 15° to the direction of motion, and a force of 30 N at an angle of θ to the direction of motion. Find the value of θ and the acceleration of the box.

6. A truck of mass 15 000 kg is being towed by two ropes. One pulls with a tension of 3000 N at an angle of 20° to the direction of motion. The other pulls with a tension, T, at an angle of 10° to the direction of motion. There is resistance of 500 N against the motion, in the same line as the motion.

   a Draw the force diagram for this situation.

   b Find T and the acceleration of the truck.

7. A ship of mass 10 000 kg is being towed due north by two tugboats with acceleration $0.1\,\mathrm{m\,s^{-2}}$. One pulls with a tension of 2000 N on a bearing of 330°. The other pulls with a tension, T, on a bearing of θ. There is resistance against the motion of 1000 N. Find T and θ.

8. A train of mass 230 tonnes provides a driving force of 300 000 N to accelerate up a slope at an angle of 5° to the horizontal. The force diagram is shown. Find the acceleration of the train.

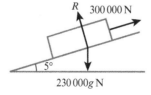

9 A log of mass 200 kg is dragged up a slope at an angle of 13° to the horizontal by a rope attached to a truck. The rope is at an angle of 20° above the slope.

   a Draw the force diagram for this situation.

   b The log accelerates at $0.3 \, \text{m s}^{-2}$. Find the tension in the rope.

10 A windsurfer and his board have a total mass of 80 kg. They are being pushed by the water with a force of 20 N westwards. The wind is pushing them northwards with a force F. The windsurfer accelerates on a bearing of 340°. Find the force F and the acceleration of the windsurfer.

11 A buoy of mass 12 kg is on the surface of a lake. The tide pushes it with a force of 25 N and the wind pushes it with a force of 15 N, as shown in the diagram. The buoy moves in the direction shown. Find the value of θ and the acceleration.

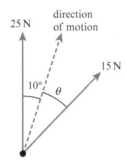

12 A girl pulls a toy car of mass 0.8 kg by a string along a horizontal path. The tension in the string is 3 N and the string is held at an angle of 40° above the horizontal. There is air resistance of 2 N. Find the time taken to reach a speed of $2 \, \text{m s}^{-1}$ from rest.

13 A shopper drags a trolley of mass 25 kg from rest along horizontal ground. The shopper is pulling the trolley by a force of 30 N with his arm, which is at 15° above the horizontal. There is friction of 10 N. Find the speed of the trolley after being pulled a distance of 6 m.

14 A ball of mass 3 kg is rolled with initial speed $4 \, \text{m s}^{-1}$ up a slope at an angle 10° to the horizontal.

   a Find the maximum distance up the slope the ball reaches.

   b What assumptions have been made to answer the question?

15 A cyclist of mass 70 kg (including her bicycle) arrives at an uphill stretch of road of length 30 m with an angle 9° to the horizontal, travelling at $10 \, \text{m s}^{-1}$. She exerts a force of 15 N parallel to the slope and there is wind resistance of 5 N against her. Find the time taken to reach the top of the slope.

16 A ball of mass m kg is rolled up a slope at an angle θ to the horizontal, where $\sin \theta = \frac{2}{5}$. The ball passes a point A with speed $7 \, \text{m s}^{-1}$. A point B is 5 m further up the slope than point A. Find the time between passing B on the way up and returning to B on the way down.

17 A van of mass 2000 kg is towed from rest by two ropes. One pulls with a tension of 130 N at 10° to the direction of motion and the other acts at 15° to the direction of motion. Find the distance covered in 10 s.

18 A ship of mass 15 000 kg is moving due east at $2 \, \text{m s}^{-1}$ when it starts being towed by a tugboat. The wind is blowing it on a bearing of 60°, so the tugboat exerts a force of 5000 N on a bearing of 100° to make the ship continue to go east. Find the speed of the ship after 5 s.

19 A box of mass 9 kg is dragged along horizontal ground by a force F acting at 30° above the horizontal. There is friction of 5 N. The box starts at rest and reaches a speed of $4\,\mathrm{m\,s^{-1}}$ in 10 m. Find the size of the force F.

 20 A car of mass 1200 kg arrives at a steep upwards slope of length 130 m at 34° to the horizontal. It is travelling at $12\,\mathrm{m\,s^{-1}}$ and there is air resistance of 100 N. Find the minimum force, assumed constant, the engine must provide for the car to reach the top of the slope.

3.5 Non-equilibrium problems and finding resultant forces and directions of acceleration

In the previous section, the direction of acceleration was known or could be worked out from the situation. In the situation here, with forces A and B, the direction of acceleration is unknown.

In situations like this, we can work out the single force equivalent to the combination of the other forces by drawing the vectors end to end, as in the following diagram. This is called the **resultant** of the other forces. If these are the only forces in the situation, the resultant is the net force for use in Newton's second law.

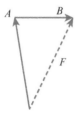

We can then use the diagram and trigonometry to work out the magnitude and direction of the resultant of the forces A and B, which is shown by F.

In the following situation with forces A, B and C, the direction of acceleration is again unknown.

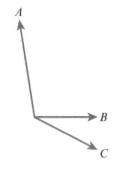

When there are three forces, if we draw the vectors end to end we will get a quadrilateral. It may not be easy to calculate the resultant from this diagram.

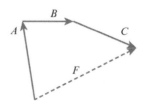

So, when there are more than two forces, we find the components of the net force by resolving horizontally and vertically. By adding these horizontal and vertical components, we can find the horizontal and vertical components, F_x and F_y, of the resultant force, F. We can use the components of the resultant to calculate the magnitude and direction of the resultant force.

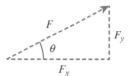

KEY POINT 3.7

The magnitude of the resultant force, F, with components F_x horizontally and F_y vertically, can be calculated using Pythagoras' theorem as $F = \sqrt{F_x^2 + F_y^2}$.

The direction of the resultant force, F, with components F_x horizontally and F_y vertically, can be calculated using trigonometry as $\tan\theta = \dfrac{F_y}{F_x}$, where θ is the angle with the x-direction.

TIP

Do not show the resultant force on the force diagram because it is easy to confuse it with a separate force. Instead, to show the resultant force, draw a second diagram alongside the force diagram.

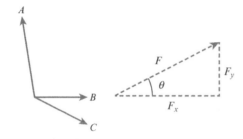

EXPLORE 3.2

Two students are discussing the following situation. A heavy stone has three ropes attached. They are pulled on bearings of 010°, 020° and 060°. Three people can pull with forces of 200 N, 150 N and 100 N. Which person should pull on which rope to maximise the net force if the direction is unimportant?

Student A	Student B
The total net force will be the same whoever pulls each rope, but the direction may change.	Who pulls each rope will affect both the net force and direction. We will need to work out each case to decide which gives the largest net force.

Which one of the students is correct?

If student A is correct, what effect does the arrangement of the people pulling the ropes have on the direction of motion and why?

If student B is correct, is there a general rule as to who should pull each rope to maximise the net force and why does it work?

If instead the direction is more important than net force, how can you decide who should pull each rope so that the net force is as close as possible to a given direction?

WORKED EXAMPLE 3.11

A boat of mass 100 kg experiences a force of 30 N eastward from the wind and a force of 40 N from the tide on a bearing of 35°, as shown in the diagram. Find the direction of the subsequent motion and the acceleration.

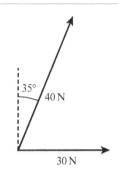

Answer

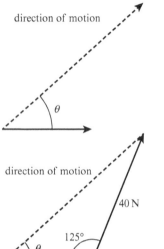

Draw a diagram with the resultant force to show where the angle is being measured from.

Adding vectors is equivalent to drawing them end to end.

Do not draw the resultant as a separate force on the force diagram.

$R^2 = 30^2 + 40^2 - 2 \times 30 \times 40 \cos 125°$
$R = 62.3 \text{ N}$

Use the cosine rule to find the resultant.

$F = ma$
$R = 100a$
$a = 0.623 \text{ m s}^{-2}$

Use Newton's second law in the direction of acceleration.

$\dfrac{40}{\sin \theta} = \dfrac{R}{\sin 125°}$

$\theta = 31.8°$

Use the sine rule to find the angle.

So the direction of motion is on a bearing of 58°.

WORKED EXAMPLE 3.12

A particle of mass 3 kg is attached to three ropes in the horizontal plane with forces of 2 N, 4 N and 3 N, as shown in the diagram.

Find the direction of the subsequent motion and the acceleration.

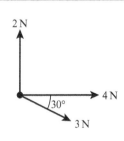

Answer

direction of motion

$F_x = 4 + 3\cos 30° = 6.60$
$F_y = 2 - 3\sin 30° = 0.50$

$\tan\theta = \dfrac{F_y}{F_x}$

$\theta = 4.33°$

so the direction is 4.33° above the positive x-direction.

$F^2 = F_x^2 + F_y^2$

$F = 6.62\,\text{N}$

$F = ma$

$F = 3a$

$a = 2.21\,\text{m s}^{-2}$

- Draw a separate diagram showing the resultant force.
- Do not draw it on the force diagram.
- Find the components of the resultant horizontally and vertically.
- Use trigonometry to find θ.
- Use Pythagoras' theorem to find the resultant.
- Use Newton's second law in the direction of acceleration.

EXERCISE 3E

1. A particle of mass 3 kg is at rest and has two forces acting on it. One has magnitude 5 N and the other has magnitude 3 N. They act in the directions shown. Find the magnitude and direction of acceleration of the resulting motion.

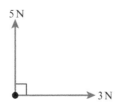

2. A mass of 4 kg is held above the ground and released from rest. There is wind blowing it with a force of 20 N horizontally. Find the angle from the downward vertical at which it initially falls.

3. A boat has its motor running, creating a force of 500 N. The wind is blowing it with a force of 200 N. The directions of the forces are shown on the diagram. Find the direction of the subsequent motion.

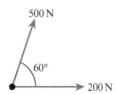

4 A particle of mass 25 kg has two forces acting on it, one of 20 N and one of 35 N, in the directions shown. Find the magnitude and direction of the resulting acceleration.

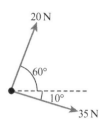

5 Three coplanar forces act on a particle, as shown in the following diagram. X has components 0 N in the x-direction and 20 N in the y-direction. Y has components 25 N in the x-direction and −10 N in the y-direction. Z has components −10 N in the x-direction and −15 N in the y-direction. Find the magnitude and direction of the resultant of the three forces.

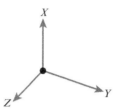

6 Three coplanar forces act on a particle, as shown in the diagram.

a Force F has components of −30 N in the x-direction and −40 N in the y-direction. Find the value of α.

b Find the magnitude and direction of the resultant of the three forces.

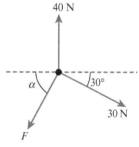

7 Four coplanar forces act on a particle, as shown in the following diagram. Show that the x component and y component of the resultant are equal. Hence, determine the direction of the resultant force.

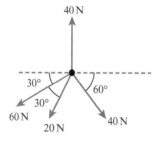

8 Three people drag a bag of sand of mass 150 kg, labelled A in the diagram. They pull in the horizontal plane with forces 40 N, 25 N and 35 N in the directions shown, compared to the direction AB.

a Find the magnitude and direction of acceleration of the resultant motion.

b What assumptions have been made to answer the question?

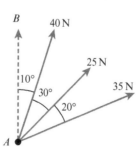

9 A boat of mass 1000 kg is pulled by three tugboats. One pulls due north with force 500 N, one pulls due east with force 350 N and one pulls on a bearing of 040° with a force of 250 N. Find the bearing and acceleration of the resultant motion.

10 In a competition of strength, four people pull a mass with ropes at different angles. The direction in which the mass moves determines the winner. Arjun wants the mass to go north, Bob wants it to go east, Chen wants it to go south and David wants it to go west. The men pull with the forces in the directions shown in the diagram. Find the direction of the resultant motion and determine who wins.

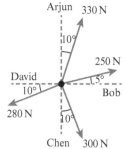

11 A rowing boat of mass 120 kg is being pulled from rest by three boats. One pulls north with a force of 100 N, the second pulls on a bearing of 020° with a force of 80 N and the third pulls on a bearing of 045° with a force of 90 N. There is resistance from the water of 200 N directly against the motion. Find the bearing and acceleration of the resultant motion.

12 A hovercraft has an engine providing a force of 150 N on a bearing of 340°. The wind blows on a bearing of 310°, which means the hovercraft accelerates from rest on a bearing of 320°. Find the force of the wind on the hovercraft.

P 13 The wind is blowing a boat with force F. The motor of the boat can exert a driving force of D N, where $D < F$. Show with a diagram that, whatever direction the wind is taking the boat with the motor switched off, the motor is capable of deflecting the direction by a maximum of $\sin^{-1}\dfrac{D}{F}$.

PS 14 A building is unstable after a natural disaster. A car is stuck under the building and needs to be dragged out as quickly as possible, although the exact direction is less important. Three people can pull ropes, one due north, one at a bearing of 010° and one at a bearing of 030°. Akhil can pull with a force of 300 N, Ben can pull with a force of 240 N and Khadijah can pull with a force of 210 N. Find who should pull each rope to maximise the acceleration and what the net force will be.

Checklist of learning and understanding

- A force can be split into components using the idea that force is a vector and can be written as the sum of other vectors.
- The components are usually found in two perpendicular directions with the force as the hypotenuse of a right-angled triangle and the other two sides as components.
- Directions chosen are usually horizontally and vertically, parallel and perpendicular to a slope, or parallel and perpendicular to the direction of motion.
- Resolving perpendicular to an unknown force means the unknown will not appear in the equation.
- When the direction of acceleration is unknown it is normally best to find components of a resultant force and use them to find the direction and magnitude of the resultant.

END-OF-CHAPTER REVIEW EXERCISE 3

1. Three forces act on a particle in equilibrium in the horizontal plane, as shown in the diagram. Find the size of the unknown force F and the angle θ.

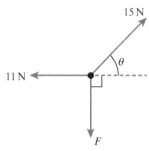

2. Three forces act on a particle in equilibrium in the horizontal plane, as shown in the diagram. By resolving in a direction perpendicular to F, show that $\theta = 47.2°$ and find F.

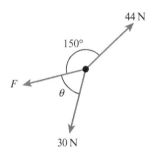

3. A girl is dragging a suitcase of mass 18 kg on horizontal ground, using a strap. The strap is at 40° to the horizontal. She pulls with a force of 15 N. There is air resistance of 5 N.

   a Find the magnitude of the normal contact force from the ground on the suitcase.

   b Find the acceleration of the suitcase.

4. Two people drag a car of mass 1200 kg forward with ropes. One pulls with force 400 N on a bearing of 005°. One pulls with force 360 N on a bearing of 352°. Find magnitude of the acceleration and its direction to the nearest 0.1°.

5. A boat is in equilibrium held by a rope to the shore. The rope exerts a force T at an angle θ from north. The wind blows the boat with force 40 N in a northwest direction. The current pushes it south with a force of 50 N. Show that $T \sin \theta = 20\sqrt{2}$ and find an expression for $T \cos \theta$. Hence, show that $\tan \theta = \dfrac{8 + 10\sqrt{2}}{17}$ and find θ and T.

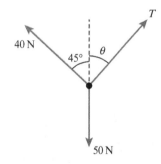

6. A car of mass 300 kg is on a slope, which is at an angle of 5° to the horizontal. When it is pulled down the slope by a rope parallel to the slope with a force of T, it accelerates at $2\,\text{m s}^{-2}$. Find the acceleration of the car when it is pulled up the slope by a rope parallel to the slope with a force of T.

PS 7. Three boys are having a strength competition. They hold ropes attached to the same object of mass 10 kg. One pulls due north with force 32 N and another pulls on a bearing of 200° with force 45 N. The third wants to make the object accelerate due east and pulls with a force of 24 N.

   a Find the bearing at which the third boy should pull.

   b Find the resultant acceleration.

8 A girl can drag a stone block of mass 18 kg up a slope at an angle of 13° to the horizontal with an acceleration of 0.7 m s^{-2}. Assuming this is the maximum force she can exert to drag the block, find the mass of the heaviest stone block she would be able to drag up the slope.

9 A box of mass 8 kg is held at rest at the top of a slope 6 m long at an angle of 12° to the horizontal. Assume air resistance and friction are negligible.

   a The box is released. Find the time taken for the box to reach the bottom of the slope.

   b Instead, a boy pushes the box downwards with a force of 20 N parallel to the slope. Find how much sooner the box reaches the bottom of the slope than under gravity alone.

10 A girl is sitting on a sledge, which her friend drags across the horizontal surface of a frozen lake. The sledge is initially at rest and then the friend pulls on a rope at an angle of 35° above the horizontal with a force of 8 N for 2 m before releasing the rope. The total mass of the girl and the sledge is 50 kg. There is air resistance of 2.4 N.

   a Find the speed of the sledge when the friend releases the rope.

   b What assumptions have been made to answer the question?

   c After the rope is released, air resistance causes the sledge to slow down until coming to rest. Find the total distance before the sledge comes to rest.

11 In a test of strength competition, a competitor must get a 10 kg stone as far as they can up a slope. The slope is at 10° to the horizontal. The competitor can drag the stone for 5 m from rest up the slope and then must release it. Frictional forces are to be considered negligible.

   a A competitor drags the stone with a rope at an angle of 16° above the slope and a force of 65 N. Find the speed at which the stone is released.

   b Find how far the stone travels after being released before coming to rest.

12 The four athletes in a bobsleigh team start the race by running along the ice. They push for 40 m on a horizontal track, providing an average horizontal force of 180 N each. The total mass of the bobsleigh and the four athletes is 600 kg.

   a Find the speed at the end of the horizontal stretch of track.

   The athletes then get into the bobsleigh. The track continues with a downhill stretch of length 1300 m on a slope at an angle of 5° to the horizontal. There is air resistance of 175 N.

   b Find the total time to complete the entire track.

13 A ball of mass m kg slides down a slope, which is at an angle of $\theta°$ to the horizontal. It passes two light gates x m apart. At the first gate, the speed of the ball is measured as u m s^{-1}, and at the second its speed is measured as v m s^{-1}. Assuming the resistance is constant, show the resistance force has a total size of $\frac{m}{2x}(2xg \sin\theta + u^2 - v^2)$.

14 A car of mass m kg is rolling down a slope of length x m, which is at an angle of 30° to the horizontal. It has a booster that provides a force of mg N over a distance of 1 m, which the driver sets off at a distance s m after the car starts moving. Assuming the booster is used before the end of the slope, show that the speed at the bottom of the slope is given by $v^2 = g(x + 2)$ and deduce that the final speed is independent of when the booster is applied. (Note that if the booster were applied for a fixed time rather than a fixed distance this would not be true.)

15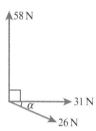

Coplanar forces of magnitudes 58 N, 31 N and 26 N act at a point in the directions shown in the diagram. Given that $\tan \alpha = \dfrac{5}{12}$, find the magnitude and direction of the resultant of the three forces. [6]

Cambridge International AS & A Level Mathematics 9709 Paper 43 Q2 November 2011

16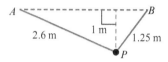

A particle P of mass 1.05 kg is attached to one end of each of two light inextensible strings, of lengths 2.6 m and 1.25 m. The other ends of the strings are attached to fixed points A and B, which are at the same horizontal level. P hangs in equilibrium at a point 1 m below the level of A and B (see diagram). Find the tensions in the strings. [6]

Cambridge International AS & A Level Mathematics 9709 Paper 43 Q3 November 2013

17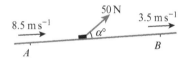

A block of mass 60 kg is pulled up a hill in the line of greatest slope by a force of magnitude 50 N acting at an angle $\alpha°$ above the hill. The block passes through points A and B with speeds $8.5 \,\text{m s}^{-1}$ and $3.5 \,\text{m s}^{-1}$ respectively (see diagram). The distance AB is 250 m and B is 17.5 m above the level of A. The resistance to motion of the block is 6 N. Find the value of α. [11]

Cambridge International AS & A Level Mathematics 9709 Paper 41 Q7 November 2014

CROSS-TOPIC REVIEW EXERCISE 1

1. A car of mass 1500 kg is on a straight horizontal road. The car accelerates from 20 m s^{-1} to 24 m s^{-1} in 10 s. The car has a constant driving force and there is resistance of 100 N. Find the size of the driving force. [4]

2. A particle starts from rest at a point X and moves in a straight line until 40 s later it reaches a point Y, which is 145 m from X. For $0\,s < t < 5\,s$ the particle accelerates at 0.8 m s^{-2}. For $5\,s < t < 30\,s$ it remains at constant velocity. For $30\,s < t < 40\,s$ it decelerates at a constant rate, but does not come to rest.

   a Find the velocity at time $t = 5\,s$ and $t = 40\,s$. [5]

   b Sketch the velocity–time graph. [2]

3. A particle P is released from rest down a slope, which is at an angle of 20° to the horizontal. There is no friction between the particle and the slope.

   a Find the particle's speed after 0.7 s. [2]

   b Find the speed when the particle has travelled 1.2 m. [2]

4. A crate of weight 400 N is lifted by a forklift truck. The truck lifts the crate from rest to a height of 2 m in 5 s. Assuming constant acceleration, find the normal contact force from the truck on the crate. [4]

5. A force F acts in a horizontal plane and has components 25 N in the x-direction and −17 N in the y-direction relative to a set of axes. The force acts at an angle α below the x-axis.

   a Find the sizes of F and α. [4]

   b Another force has magnitude 29 N and acts at an angle of 70° above the positive x-axis. The resultant of these two forces has magnitude R N and makes an angle of θ with the positive x-axis. Find the values of R and θ. [3]

6. The graph shows the velocity of a parachutist as she falls from an aircraft until she hits the ground 50 s later.

   There are four stages to the motion: falling freely under gravity with the parachute closed; decelerating with the parachute open; falling at constant speed with the parachute open; and coming to rest instantaneously on hitting the ground.

   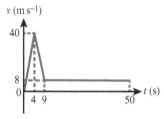

   a Find the total distance fallen. [2]

   b The parachutist has mass 70 kg. Show that the upward force on the parachutist due to the parachute during the second stage is 1148 N. [5]

7. A particle of mass 6.3 kg is attached to one end of a light inextensible string. The other end of the string is attached to a fixed point X. It is held in equilibrium by a horizontal force F when the string is at an angle α to the vertical, where $\tan\alpha = \dfrac{20}{21}$. Find the tension in the string and the size of F. [4]

8. Two forces, each of size 8 N, have a resultant of 13 N.

   a Find the angle between the forces. [2]

   b The two given forces of magnitude 8 N act on a particle of mass m kg, which remains at rest on a horizontal surface with no friction. The normal contact force between the surface and the particle has magnitude 7 N. Find m and the acute angle that one of the 8 N forces makes with the surface. [3]

9   Three coplanar forces of magnitudes 7 N, 10 N and 12 N act at a point A, as shown in the diagram.

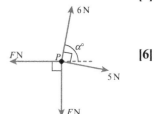

   a   Find the component of the resultant of the three forces in the direction AB and perpendicular to the direction AB. [3]

   b   Hence, find the magnitude and direction of the resultant of the three forces. [3]

10 a   A cyclist lets her bike accelerate down a slope with constant gradient, at constant acceleration. She passes a point A, then 4 s later passes a point B 32 m away. Another 2 s later she passes a point C a further 19 m away. Find the acceleration of the cyclist. [5]

   b   Assuming there is no friction or resistance and the cyclist is not pedalling, find the angle that the slope makes with the horizontal, giving your answer to the nearest 0.1°. [3]

11 A particle P is in equilibrium on a smooth horizontal table under the action of four horizontal forces of magnitudes 6 N, 5 N, F N and F N acting in the directions shown. Find the values of α and F. [6]

Cambridge International AS & A Level Mathematics 9709 Paper 42 Q3 November 2010

12 A cyclist starts from rest at point A and moves in a straight line with acceleration $0.5\,\text{m s}^{-2}$ for a distance of 36 m. The cyclist then travels at constant speed for 25 s before slowing down, with constant deceleration, to come to rest at point B. The distance AB is 210 m.

   i   Find the total time that the cyclist takes to travel from A to B. [5]

   24 s after the cyclist leaves point A, a car starts from rest from point A, with constant acceleration $4\,\text{m s}^{-2}$, towards B. It is given that the car overtakes the cyclist while the cyclist is moving with constant speed.

   ii Find the time that it takes from when the cyclist starts until the car overtakes her. [5]

Cambridge International AS & A Level Mathematics 9709 Paper 41 Q7 November 2015

13 A small bead Q can move freely along a smooth horizontal straight wire AB of length 3 m. Three horizontal forces of magnitudes F N, 10 N and 20 N act on the bead in the directions shown in the diagram. The magnitude of the resultant of the three forces is R N in the direction shown in the diagram.

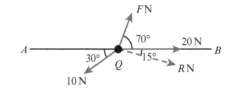

   i   Find the values of F and R. [5]

   ii Initially the bead is at rest at A. It reaches B with a speed of $11.7\,\text{m s}^{-1}$. Find the mass of the bead. [3]

Cambridge International AS & A Level Mathematics 9709 Paper 41 Q5 November 2015

14 A particle P of weight 21 N is attached to one end of each of two light inextensible strings, S_1 and S_2, of lengths 0.52 m and 0.25 m respectively. The other end of S_1 is attached to a fixed point A, and the other end of S_2 is attached to a fixed point B at the same horizontal level as A. The particle P hangs in equilibrium at a point 0.2 m below the level of AB with both strings taut (see diagram). Find the tension in S_1 and the tension in S_2. [6]

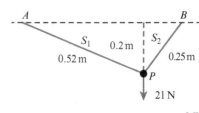

Cambridge International AS & A Level Mathematics 9709 Paper 43 Q4 November 2012

Chapter 4
Friction

In this chapter you will learn how to:

- calculate the size of frictional forces
- use friction to solve problems in motion
- determine the direction of motion of an object
- solve problems where a change in direction of motion changes the direction of friction.

Cambridge International AS & A Level Mathematics: Mechanics

PREREQUISITE KNOWLEDGE

Where it comes from	What you should be able to do	Check your skills
IGCSE / O Level Mathematics	Use Pythagoras' theorem.	1 Find the hypotenuse of a right-angled triangle with short sides of length 8 m and 15 m.
IGCSE / O Level Mathematics	Use trigonometry for right-angled triangles.	2 A triangle, $\angle ABC$, has a right angle at B. Length BC is 7 m and $\angle BAC$ is 35°. Find length AC.
Chapter 2 / Chapter 3	Resolve forces and use Newton's second law.	3 A box of mass 5 kg is on a slope at an angle of 10° to the horizontal. It is pulled down a slope with force 8 N parallel to the slope. Find the acceleration of the box.

How does friction work?

When a box is at rest on the floor it is in equilibrium with the weight balanced by a contact force. It does not matter if there is friction or not because no frictional force is required for the box to stay in equilibrium. However, when you gently push the box horizontally, it may still remain in equilibrium and not move. This is because friction prevents it. As you increase the pushing force the box may still not move. This suggests that friction can change value in order to prevent motion.

At some point the pushing force on the box will be large enough to overcome friction and the box will slide along the floor. What factors affect the point at which this occurs? Does it depend on the size or shape of the object? It is reasonable to expect the size of the force will depend on the two surfaces in contact. But what else affects it?

Once the force is large enough to overcome friction, how does friction behave? Does friction remain fixed or does it change depending on the motion?

All these questions will be considered in this chapter.

4.1 Friction as part of the contact force

EXPLORE 4.1

Connect a spring balance to a block of wood on a horizontal surface. Increase the force on the spring balance horizontally until the block starts moving. When the block is at rest, the friction force takes a large enough value to prevent motion. When it starts moving, try to keep it moving slowly at a constant speed and read off the force on the spring balance. This will be equivalent to the frictional force. Try this on different surfaces and you should see that some surfaces have different amounts of friction. Try moving the block at different constant speeds. The size of friction should not be affected by the speed of the object.

Try resting a small mass on top of the block before pulling it horizontally. The frictional force should be larger now. This suggests that the mass may affect the size of friction. However, the mass also affects the normal contact force. By adding a mass and simultaneously lifting the block slightly with another spring balance (quite difficult in practice), the size of the force of friction goes down again despite the larger mass. This suggests it is the size of the normal contact force that affects friction, not the mass.

If there is friction between two surfaces, the contact is called rough. If there is no friction the contact is called smooth.

If there is no motion, friction takes whatever value is required to prevent motion. This means that if an object is at rest on a horizontal surface with no forces other than its weight acting on it, there will be no friction. When a force acts on the object, but is not strong enough to cause motion, friction will act in the opposite direction to the force. As the force is increased friction will increase until the point when the force is large enough to overcome the friction and cause motion.

When the force on the object is large enough that the object is still in equilibrium but any more force would cause motion, the object is said to be in limiting equilibrium. At this point friction will take a fixed, maximum value. That value depends on two main factors: how rough the surfaces are and the normal contact force between them. Each pair of surfaces has a coefficient of friction, denoted by μ, which gives a numerical value for how rough the surface is. The size of friction is limited to a value μ times the normal contact force.

DID YOU KNOW?

Surprisingly, friction will not noticeably depend on the amount of area in contact between the two surfaces. A larger area would create more friction, but it also spreads out the normal contact force over a larger area so has almost no net effect.

KEY POINT 4.1

Friction can take any value up to its limiting value.
$$F \leqslant \mu R$$
If the object is moving relative to the surface, friction will take the limiting value:
$$F = \mu R$$
where R is the normal contact force.

A typical value for μ is between 0.3 and 0.7, although surfaces that are not as rough may have a smaller coefficient of friction and surfaces that are extremely rough may have a larger coefficient of friction, possibly bigger than 1, although this is unusual. A 'smooth' surface has no friction, which is equivalent to μ and takes the value 0.

Friction depends on the normal contact force between the surfaces, so we say it is 'part of the contact force', and it acts parallel to the surface whereas the normal contact force is perpendicular. This means the total contact force is the resultant of the normal contact force and the friction force. We calculate it in the usual way by considering a right-angled triangle and using Pythagoras' theorem.

KEY POINT 4.2

The total contact force can be found from $C = \sqrt{R^2 + F^2}$.

The direction of the total contact force is at an angle of $\tan^{-1}\left(\dfrac{F}{R}\right)$ to the normal contact force.

When an object is stationary, but about to move, friction will take the limiting value. We say the object is in limiting equilibrium, or we can use the phrase 'on the point of slipping'. This means that any extra force would make the object start moving.

Sometimes it is not clear which way friction acts. Suppose a car is on a rough slope with a tow rope attached on the end of the car facing up the slope. Tension in the rope is acting parallel to the **line of greatest slope** in an upwards direction. We don't know if the tension is there to try to pull the car up the slope, or to help prevent the car moving down the slope. Friction will act in different directions depending on the situation.

The two possibilities are shown by the two force diagrams. If the tension in the rope is large, friction may act down the slope to prevent the car going up the slope (shown in the left diagram). If the tension is small, friction may act up the slope to prevent the car going down the slope (shown in the right diagram).

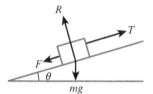
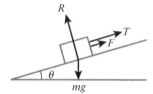

In situations where it is not clear which way friction acts, you must make an assumption about what happens in the subsequent motion. You need to be aware of the significance of getting a negative value.

KEY POINT 4.3

If you assume friction acts in one direction and then solving the equations gives a negative value for friction, it means your assumption was wrong. It means that friction has the same magnitude but in the other direction. You should state that the direction is not as marked on your diagram.

The value of friction given by equations may be too large because friction is limited to μR. If that happens, then the object cannot be in equilibrium.

KEY POINT 4.4

If the value calculated for friction to keep the object in equilibrium is larger than the limiting value, the object cannot remain in equilibrium.

TIP

Forces 'parallel to the slope' act along the line of greatest slope. Any other direction parallel to the surface will not be as steep, which is why roads up steep slopes wind up rather than go straight up. In this course, forces will generally act along lines of greatest slope.

WORKED EXAMPLE 4.1

A box of mass 5 kg is at rest on horizontal ground. The box is being pulled by a horizontal force of 8 N. Find the total contact force.

Answer

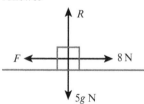

$R = 5g = 50 \text{ N}$ Resolve vertically to find R.

$F = 8$ N

Resolve horizontally to find F.

Note that the box is at rest, so friction must be of the magnitude to prevent motion.

$C = \sqrt{F^2 + R^2} = 50.6$ N

Use Pythagoras' theorem to find the total contact force.

WORKED EXAMPLE 4.2

a A book of mass 8 kg is at rest on a rough slope, which is at an angle of 20° to the horizontal. The book is held in limiting equilibrium by a force of 10 N up the line of greatest slope. Find the coefficient of friction and the magnitude of the contact force.

b Find the largest force up the slope for which the book remains at rest.

Answer

a

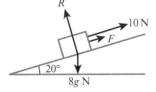

The force diagram assumes the book is on the point of slipping down rather than up the slope.

Note that it may not be obvious from the question whether the book is on the point of slipping up or down the slope.

$R = 8g \cos 20 = 75.2$ N

Resolve perpendicular to the slope first to find R.

$8g \sin 20 = F + 10$
$F = 17.4$ N

Resolve parallel to the slope to find F.

If you had assumed the book was about to slip up the slope and marked friction as acting down the slope, you would have got −17.4 N for friction and realised you had made the wrong assumption: friction should be the other way.

$F = \mu R$
$\mu = \dfrac{F}{R}$
$= 0.231$

The book is on the point of slipping, so friction is limiting.

$C = \sqrt{F^2 + R^2}$
$= 77.2$ N

Use Pythagoras' theorem to find the contact force.

b

If the force up the slope is the largest possible to prevent motion, friction must act down the slope.

$R = 8g \cos 20$ Resolve perpendicular to the slope first to find R as before.
$= 75.2$ N

$8g \sin 20 + F = P$ Resolve parallel to the slope to find P.

$F = \mu R$ Since the book is on the point of slipping, friction is limiting
$= 17.4$ N   so we can find F using the value for μ found in part **a**.

$P = 44.7$ N

WORKED EXAMPLE 4.3

A waste container of mass 400 kg is in equilibrium on a rough slope at an angle of 18° to the horizontal. The coefficient of friction between the slope and the skip is 0.3. It is held in equilibrium by a winch with tension T N. Find the range of possible values for T.

Answer

When T is minimal:

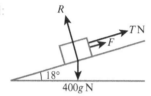

Firstly, consider the case where the winch is providing the minimum force to prevent the skip from sliding down the slope.

$R = 400g \cos 18 = 3800$ N   Resolve perpendicular to the slope first to find R.

$400g \sin 18 = T + F$ Resolve parallel to the slope next to find T.

$F = \mu R = 1140$ Since the tension is the minimum possible, friction must
$T = 94.8$ take the maximum value.

When T is maximal:

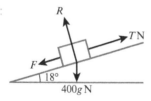

Secondly, consider the case where the winch is providing the maximum force, which is not enough to make the skip slide up the slope, so friction acts down the slope in this case.

$R = 400g \cos 18 = 3800$ N   Resolve perpendicular to the slope first to find R as before.

$400g \sin 18 + F = T$ Resolve parallel to the slope next to find T.

$F = \mu R = 1140$ Since the tension is the maximum possible, friction must
$T = 2380$ take the maximum value.

Therefore, the range of values for T is $94.8 \leq T \leq 2380$.

Chapter 4: Friction

EXERCISE 4A

1 A box is at rest on horizontal ground.

   a When it is pulled to the right by a force of 40 N, as shown in the diagram, find the size and direction of the force of friction.

   b When it is instead pushed to the left by a force of 25 N at 20° above the horizontal, find the size and direction of the force of friction.

   c When there is no sideways force acting on the box, find the size of the force of friction.

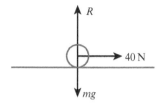

2 A box of mass 14 kg is at rest on a slope that is at 15° to the horizontal.

   a When there is no external force acting on the box, find the size and direction of the force of friction.

   b When it is pulled up the slope by a force of 50 N parallel to the line of greatest slope, as shown in the diagram, find the size and direction of the force of friction. (Note that friction is not marked. You will have to decide which direction you think friction is acting and come to a conclusion based on whether the answer you get is positive or negative.)

   c When the box is dragged down the slope by a force of 20 N at 10° above the line of greatest slope, find the size and direction of the force of friction.

   d When it is pulled up the slope by a force of 15 N at 35° above the horizontal, find the size and direction of the force of friction.

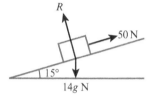

3 A box of mass 20 kg is at rest on rough horizontal ground. Find the magnitude of the total contact force in each of these cases.

   a The box is pulled horizontally to the right by a force of 40 N.

   b The box is pushed to the left by a force of 50 N at 15° above the horizontal, as shown in the diagram.

   c The box is pushed to the left by a force of 50 N at 15° below the horizontal.

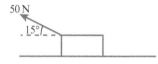

4 A book of mass 4 kg is at rest on a rough slope at angle 14° to the horizontal. Find the magnitude of the total contact force in each of these cases.

   a No other force acts on the book.

   b The book is pulled down the slope by a force of 5 N parallel to the line of greatest slope.

   c The book is pulled up the slope by a force of 15 N at 9° above the line of greatest slope, as shown in the diagram.

5 A tin of mass 0.5 kg is on a rough horizontal table with coefficient of friction 0.3. Find the largest horizontal force that can be exerted on the tin before the tin starts to move.

6 A block of wood of mass 3 kg is on a rough slope, which is at an angle of 25° to the horizontal. The coefficient of friction between the block and the slope is 0.4. It is held in place by a force, P, going up the line of greatest slope.

   a Find the smallest possible size of P to prevent the block sliding down the slope.

   b Given that the block remains in equilibrium, find the largest possible size of P.

7 A chair of mass 6 kg is at rest on a rough horizontal floor with coefficient of friction 0.35. It is pulled horizontally by a force of 25 N. A boy pushes down on the chair so that the chair is on the point of slipping but remains at rest. Find the force that the boy exerts on the chair.

8 Two men are trying to drag a bin of mass 100 kg up a rough slope at an angle 20° to the horizontal. The coefficient of friction is 0.25. One man pulls up the slope with a force of 400 N. The other tries to lift the bin perpendicularly to the slope, providing a force such that the bin is on the point of slipping up the slope. Find the force exerted by the second man.

9 A sledge of mass 200 kg is being pulled by a woman along rough horizontal ground. She exerts a force of 500 N at 18° above the horizontal and the sledge is on the point of slipping. Find the coefficient of friction.

10 A gardener is trying to move a heavy roller of mass 150 kg along rough ground at an angle of 5° to the horizontal. He exerts a force of 200 N down the slope and parallel to it and the roller is on the point of slipping.

   a Find the coefficient of friction.

   b What assumptions have been made to answer the question?

11 In a factory, a machine picks up a box by clamping it on both sides. The box of mass 4 kg is held clamped on both sides by identical clamps with the contacts horizontal. The machine provides a contact force of 50 N with each clamp. Find the minimum coefficient of friction between each clamp and the box for the box not to slip.

12 A box of mass 30 kg is at rest on a rough slope at an angle of 20° to the horizontal. When a girl pushes up the slope along the line of greatest slope with a force of 25 N, the box does not slip down. Find the range of values for the coefficient of friction between the box and the slope.

13 A ring of mass 2.5 kg is threaded on to a fixed horizontal wire. It is made of a rubbery material to give it an extremely high coefficient of friction (above 1) and prevent it sliding along the wire. When it is at rest, the higher part of the ring is in contact with the wire, so the normal contact force from the wire is upwards, as shown in the diagram. The ring is attached to a string, which provides a tension of 60 N at an angle of 50° above the horizontal. The ring is now in limiting equilibrium. The force diagram for the situation is given in the diagram. Note that the normal contact force is now acting downwards because there cannot be a vertical component of acceleration, so the lower part of the ring is now in contact with the wire. Find the coefficient of friction between the ring and the wire.

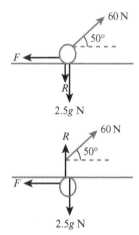

14 A box of mass 50 kg is at rest on a slope, which is at an angle of 26° to the horizontal. The coefficient of friction is 0.4. The box is held in place by a rope attached to a winch pulling up the slope and parallel to it. Find the minimum and maximum possible values for the tension, T, which the winch could provide for the box to remain in equilibrium.

15 A car of mass 1350 kg is at rest on a rough slope at an angle of 7° to the horizontal. A man tries to push it down the slope, exerting a force of 500 N, but cannot get it to move.

   a Find the angle that the total contact force makes with the slope.

   b When the man stops pushing, the car remains in equilibrium. Find the angle that the total contact force makes with the slope.

16 A ring of mass 2 kg is held in place at rest on a rough horizontal wire. It is attached to a string that is at an angle of 40° above the horizontal.

   a Explain why once the ring is released it can never be in equilibrium, however high the coefficient of friction, when the tension in the string satisfies $T \sin 40 = 2g$.

   b Show that when the tension is 100 N the coefficient of friction must be at least 1.73 for the ring to be in equilibrium, but when the tension increases to 200 N the coefficient of friction can be as low as 1.41 with the ring remaining in equilibrium. Explain why.

17 A ring of mass 3 kg is at rest on a rough horizontal wire. It is attached to a string that is at an angle of 60° above the horizontal. The coefficient of friction between the ring and the wire is 0.7. Find the set of values for the tension, T, which will allow the ring to remain in equilibrium.

4.2 Limit of friction

We have seen that when an object is in limiting equilibrium or on the point of slipping, friction takes the maximum value. When an object is moving, friction will remain at the limiting value.

KEY POINT 4.5

If the object is moving relative to the surface, friction will take the value $F = \mu R$.

When an object is moving or about to start moving, mark the friction as μR on the force diagram.

When an object moves at constant speed it is in equilibrium. However, when an object on a surface is accelerating, it will accelerate parallel to the surface. On horizontal ground the acceleration will be horizontal. On a slope, the acceleration will be along the line of greatest slope.

If we resolve parallel to the surface to find acceleration, we will not find a solution because the size of friction is not known. The size of friction will depend on two factors, μ, which may be given, and R. We will normally need to resolve perpendicular to the surface where there is no acceleration to calculate the normal contact force first. This will allow us to find the value of R and, hence, friction. Then we can resolve parallel to the surface using Newton's second law to find acceleration.

KEY POINT 4.6

In order to calculate the acceleration in the direction parallel to a rough surface, resolve perpendicular to the surface first to find the normal contact force and, hence, the frictional force. Then resolve parallel to the surface and calculate acceleration using $F = ma$.

EXPLORE 4.2

Two students, Basma and Bijal, are discussing the best way to drag a heavy box along a rough horizontal surface. Here are their arguments.

Basma	Bijal
I would pull horizontally to get all the force I can exert on the box working in the direction I want to go.	I would pull at an angle above the horizontal. This would reduce the contact force and therefore reduce the friction.

Discuss which argument is more convincing.

Practical experiments may help you answer the question. Test the situation using a wooden block and spring balance. Increase the horizontal force until it is just less than the force required to start the block moving. Try to keep the force the same, but change the angle at which it acts. Does the block start moving if the force is acting at an angle?

In Section 4.1 we considered the situation where a car is held on a slope, but we didn't know which way friction was acting. You also need to know how to deal with situations where it is not known if there is motion nor, if there is, in which direction the motion would be. Start by assuming the situation that seems likely to be correct, but be ready to spot a contradiction.

Consider the same example where a car is on a rough slope and there is a rope pulling up the line of greatest slope, but this time we do not know whether the car remains stationary.

If we assume the car slips down the slope, the friction must be limiting and act up the slope. However, if we solve the equations and get a negative value for acceleration, this contradicts the assumption and suggests the car does not, in fact, slide down the slope.

If instead we assume the car is pulled up the slope, the friction must be limiting and act down the slope. However, if this leads to a negative value for acceleration, this again would contradict the assumption and suggests the car is not, in fact, pulled up the slope.

These two results together would lead to the conclusion that the car is in equilibrium and friction may not be limiting.

 WEB LINK

You may want to have a go at the resource *A frictional story* at the *Vector Geometry* station on the Underground Mathematics website.

 KEY POINT 4.7

It may be necessary to make an assumption about the direction of motion when setting up the force diagram. If the outcome contradicts the assumption, then you need to change your initial assumption.

MODELLING ASSUMPTIONS

We have assumed that the limiting value for friction is the same whether the object is moving or not. In reality, there is a small difference between static friction and dynamic friction. From the experiment in Explore 4.2 you may have realised that to start the block moving takes slightly more force than the amount required to keep it at constant velocity once it is already moving. The difference is slight and for the purposes of this course we will ignore it and assume they are both the same.

Once the object moves, the exact point on the surface in contact with the object is always changing, so each part of the contact may have a different value for the coefficient of friction. We will assume that the difference in the values of μ across a broadly similar surface is negligible. If the surface changes significantly, this will be stated in the question and we will use a different value for μ for the different surface.

Awkward shapes may make it difficult for an object to slide smoothly along a surface. For example, a hook shape may lodge itself in the surface. However, in this course we are treating objects as particles so, whatever the size and shape of the actual object, the size of friction will not be affected by those factors.

Chapter 4: Friction

> **DID YOU KNOW?**
>
> Frederick the Great, King of Prussia from 1740 until 1786, wanted to build a fountain 30 m tall for his gardens at Sanssouci. He asked Leonhard Euler (1707–1783), one of the greatest mathematicians of the age, to help calculate how to get the water from the river under enough pressure to create the fountain. Euler did his calculations assuming no friction, but advised the engineers that he would need to do experiments to see if the calculations were valid.
>
> The engineers did not take his advice and the fountains were built according to theory alone. The pipes burst and the water never made it to the fountain. Frederick blamed Euler, despite Euler's warnings.
>
> Euler was the first to create equations modelling frictionless fluids, but it took more than a century to work out how to add friction to the model of fluid dynamics in equations known as the Navier–Stokes equations. These are still not fully understood and there is a $1 million prize for solving other aspects of these equations.

WORKED EXAMPLE 4.4

a A curling player tries to slide a curling stone of mass 20 kg along a horizontal ice rink to stop on top of a target that is 46 m away from where it was released. The coefficient of friction between the ice and the stone is 0.05. The player releases the stone with a speed of $6.5 \, \text{m s}^{-1}$. Find how far from the target it stops.

b In a game of curling there are sweepers who sweep the ice to polish it and reduce the coefficient of friction. Assuming they lower the coefficient equally along the entire path, find the reduced coefficient of friction required to get the stone to land on the target.

Answer

a

	It is useful to add the direction of motion to the diagram and show the acceleration in that direction, even though the acceleration will be negative.
$R = 20g$ $ = 200 \, \text{N}$	Resolve vertically first to find R.
$F = ma$ $-\mu R = 20a$ $a = -0.5 \, \text{m s}^{-2}$	Resolve horizontally to find a.
$v^2 = u^2 + 2as$ $0^2 = 6.5^2 - 2 \times -0.5s$ $s = 42.25$	Use an equation of motion for constant acceleration to find the distance.
Distance from target $= 46 - s$ $\phantom{\text{Distance from target}} = 3.75 \, \text{m}$	Make sure you answer the question.

b $v^2 = u^2 + 2as$

$0^2 = 6.5^2 - 2a \times 46$

$a = -0.459\,\text{m s}^{-2}$

As before,

$R = 20g$

$= 200$

and

$-\mu R = 20a$

$\mu = 0.0459$

Use an equation of motion for constant acceleration to find the acceleration.

Solve the equations to find μ.

WORKED EXAMPLE 4.5

A woman drags a box of mass 20 kg up a rough slope. The slope is at an angle of 10° to the horizontal and the coefficient of friction between the box and the slope is 0.45. The woman pulls the box using a rope held at an angle of 20° above the slope, with a tension of 120 N. Find whether the force is large enough to create motion and, if it is, find the acceleration.

Answer

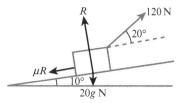

$R + 120 \sin 20 = 20g \cos 10$

$R = 156\,\text{N}$

$F = ma$

$120 \cos 20 - 20g \sin 10 - \mu R = 20a$

$a = 0.393\,\text{m s}^{-2}$

Draw a force diagram, making the 20° angle with a dotted line parallel to the slope.

This diagram assumes that there will be motion up the slope, so friction will be limiting down the slope.

Resolve perpendicular first to find R.

Resolve parallel to the slope to find a.

This is positive, so consistent with the assumption that there is motion up the slope.

EXERCISE 4B

1 A box of mass 14 kg is on horizontal ground. It is dragged by a horizontal force of 21 N. The surface is rough and the coefficient of friction between the surface and the box is 0.1.

   a Resolve vertically to find the size of the normal contact force.

   b Find the size of the frictional force.

   c Find the acceleration of the box.

2   A skip of mass 3000 kg is held at rest by a winch on a slope at an angle of 15° to the horizontal. The slope is rough and the coefficient of friction between the slope and the skip is 0.25. When the winch is removed the skip starts to slide down the slope.

   a Resolve perpendicular to the slope to find the size of the normal contact force.

   b Find the size of the frictional force.

   c Find the acceleration of the skip.

3   A boy is dragging a box of mass 20 kg up a rough slope at an angle of 12° to the horizontal. The coefficient of friction is 0.28. He provides a force of 100 N parallel to the slope. Find the acceleration of the box.

4   A gardener is pulling a wheelbarrow of mass 8 kg from rest along rough horizontal ground. The coefficient of friction between the wheelbarrow and the ground is 0.6. The gardener provides a force of 50 N at an angle of 30° above the horizontal, as shown in the diagram.

   a Find the acceleration of the wheelbarrow.

   b What happens when the wheelbarrow has 20 kg of soil in it and the gardener exerts the same force at the same angle?

5   A ski-plane has skis to land and take-off on snow. It has a mass of 3000 kg and has a propeller providing a force of 20 000 N horizontally. It accelerates from rest on horizontal ground at 2.2 m s^{-2}. Find the coefficient of friction between the ground and the ski-plane.

6   A bin of mass 16 kg is held on a downhill slope at an angle of 20°. When the bin is released, it slides down the slope with acceleration 1.2 m s^{-2}. Find the coefficient of friction between the bin and the ground.

7   A ring of mass 2 kg is on a fixed rough horizontal wire with coefficient of friction 0.4. It is pulled by a rope with tension 15 N at an angle of 5° above the horizontal. Find the acceleration of the ring.

8   A ring of mass 3 kg is on a fixed rough horizontal wire. It is pulled by a rope with tension 20 N at an angle of 10° above the horizontal and accelerates at 2 m s^{-2}. Find the coefficient of friction between the ring and the wire.

9   A ski-plane of mass 5000 kg accelerates from rest along a rough horizontal runway of length 600 m. It needs to reach a speed of 25 m s^{-1} by the end of the runway to take off. The propeller provides a horizontal force of 16 000 N. Find the maximum coefficient of friction to allow the ski plane to take off.

10 A downhill skier of mass 80 kg is accelerating down a rough slope of length 400 m at 22° to the horizontal. There is air resistance of 50 N and the coefficient of friction between the snow and the skis is 0.3. The skier is moving at 20 m s^{-1} at the top of the slope. Find the speed of the skier at the bottom of the slope.

11 A bag of sand of mass 200 kg is being winched up a slope, which is at an angle of 6° to the horizontal. The slope is rough and the coefficient of friction is 0.4. The winch provides a force of 1000 N parallel to the slope. At the bottom of the slope the bag is moving at 2 m s^{-1}. Find the distance it has moved when its speed has reduced to 1.5 m s^{-1}.

12 A man wants to drag a block of wood of mass 50 kg along horizontal rough ground, where the coefficient of friction is 0.45. If he pushes it he can generate a force of 250 N horizontally. Alternatively, he can pull via a string with a force of only 230 N at an angle of 25° above the horizontal. Which would give the larger acceleration?

13 Two men are pushing a palette of bricks of mass 120 kg along rough horizontal ground. The first man pushes horizontally with a force of 150 N. The second man pulls via a rope at an angle of 20° above the horizontal with a force of 140 N. They maintain a constant velocity.

   a Find the coefficient of friction between the palette and the ground.

   b The second man no longer pulls the rope. By first finding the new normal contact force, find the deceleration of the palette.

14 A snooker ball of mass 0.4 kg is struck towards a cushion from 0.8 m away with speed $3\,\text{m s}^{-1}$. The surface of the snooker table has a coefficient of friction of 0.3. When the ball bounces from the cushion its speed is reduced by 20%. Find how far from the cushion it stops.

15 A wooden block of mass 10 kg is on rough horizontal ground with coefficient of friction 0.6. It is dragged by a force of 80 N acting at 15° to the horizontal.

   a Find the acceleration if the force is above the horizontal.

   b Find the acceleration if the force is below the horizontal.

16 A box of mass 50 kg is slowing down from $10\,\text{m s}^{-1}$ on rough horizontal ground. The coefficient of friction between the box and the ground is 0.3. To start with, the box is being slowed by a string providing a tension of 25 N horizontally. Then the string breaks and the box comes to a halt under friction alone after a total distance of 14.5 m.

   a Find how far the box travelled before the string broke.

   b What assumptions have been made to answer the question?

4.3 Change of direction of friction in different stages of motion

A shopper is pushing a shopping trolley, but rather than just pushing it, the shopper gives it a shove, lets go and walks after it. After a few metres, the trolley stops because of friction.

When the shopper does the same thing up a slope, friction also causes the trolley to stop, but once the trolley has stopped, friction then acts in the opposite direction to prevent the trolley falling back down the slope.

When the shopper does the same thing up a steeper slope, the trolley may start moving back towards the shopper. In this situation, friction will be limiting to start with and act down the slope to stop the trolley moving up the slope. Once the trolley comes to rest, friction will act up the slope to try to prevent the trolley moving back down the slope. If the force due to gravity is large enough, the trolley will start moving back down the slope and friction will again become limiting, but will now act up the slope.

KEY POINT 4.8

When the motion of an object can be split into different stages, you need to draw a different force diagram for each stage and deal with the stages separately. The direction of the frictional force will be different if the object changes direction.

Chapter 4: Friction

EXPLORE 4.3

Two students, Nina and Jon, are discussing the problem of a ball rolling up a slope and then back down the slope.

Nina says she can save a lot of time in working out how long it takes to return to the starting point, by working out how long it takes to reach the highest point and doubling it. She says the speed when it reaches the starting point on the way down will be the same as when it started on the way up.

Jon says that's not true. The uphill stage and downhill stage have to be worked out separately. He says that the downhill bit will take longer and the speed will be lower because friction has slowed down the ball.

Nina says that's nonsense. Of course friction will slow it down more quickly when going uphill, but that just means it stops after a smaller distance and a smaller time than without friction. It will still return to the starting point after twice the time it took to reach the highest point.

Who is correct?

MODELLING ASSUMPTIONS

When an object is placed on a slope, it may topple over rather than slide down the slope. However, for this course, we are considering all objects as particles, so they have no shape and cannot fall over.

When a ball is placed on a slope, the centre of the ball is always above a point lower on the slope than the point where the ball touches the slope. Therefore, the ball will always roll down the slope, regardless of how much friction there is. Rolling is also different from sliding. However, because we are considering all objects as particles, objects like balls or cylinders, which may roll, are treated as particles that are sliding and we will ignore any differences this might give.

WORKED EXAMPLE 4.6

a A box of mass 10 kg is pushed from rest along rough horizontal ground by a horizontal force of size 50 N for 3s. The coefficient of friction is 0.45. Find the speed when it stops being pushed.

b The box then slows down because of the friction. Find the total distance the box has moved.

Answer

a

Draw the force diagram with friction limiting because we know the box will move.

$R = 10g$
$= 100$ N
$F = ma$
$50 - \mu R = 10a$
$a = 0.5$ m s^{-2}

$v = u + at$
$= 0 + 0.5 \times 3$
$= 1.5$ m s^{-1}

Resolve vertically first to find R.

Resolve horizontally, taking the direction of motion as positive, to find a.

Use an equation of constant acceleration to find the velocity.

b $\quad s = ut + \dfrac{1}{2}at^2$
So $s_1 = 0 \times 3 + \dfrac{1}{2} \times 0.5 \times 3^2$
$= 2.25$

Use an equation of constant acceleration to find the displacement for the first stage.

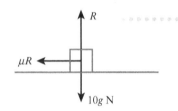

Draw a new force diagram for the second stage of the motion because the situation has changed.

$R = 10g = 100$ N

Resolve vertically to find the new value for R, which in this case is the same as the old value.

$-\mu R = 10a$
$a = -4.5$ m s^{-2}

Resolve horizontally, taking the direction of motion as positive, to find the value for a for the second stage.

$v^2 = u^2 + 2as$
So $0^2 = 1.5^2 + 2 \times -4.5 s_2$
$s_2 = 0.25$

Use an equation of constant acceleration to find the displacement for the second stage.

Hence $s = s_1 + s_2$
$= 2.25 + 0.25$
$= 2.5$ m

Find the total distance for the two stages of the motion.

> **REWIND**
>
> Look back to Chapter 1, Section 1.3, if you need a reminder of the equations of constant acceleration.

WORKED EXAMPLE 4.7

a A ball of mass 3 kg rolls up a slope with initial speed 10 m s^{-1}. The slope is at an angle of 20° to the horizontal and the coefficient of friction is 0.3. By modelling the ball as a particle, find the distance up the slope when the ball comes to rest.

b Show that after coming to rest the ball starts to roll down the slope.

c Find the speed of the ball when it returns to its starting point.

Answer

a

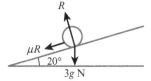

$R = 3g \cos 20$
$ = 28.2\,\text{N}$

$F = ma$
$-\mu R - 3g \sin 20 = 3a$
$a = -6.24\,\text{m s}^{-2}$

$v^2 = u^2 + 2as$
$0^2 = 10^2 - 2 \times 6.24 \times s$
$s = 8.01\,\text{m}$

b

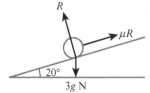

$R = 3g \cos 20$
$ = 28.2\,\text{N}$

$3g \sin 20 - \mu R = 3a$
$a = 0.601\,\text{m s}^{-2}$

This is positive, so the ball will roll back down the slope.

Alternatively, suppose the ball is in equilibrium.
Then, $F = 3g \sin 20$
$ = 10.3$
But $\mu R = 8.46 < F$, which is not possible.

c $\quad v^2 = u^2 + 2as$
$ = 0^2 + 2 \times 0.601 \times 8.01$
$v = 3.10\,\text{m s}^{-1}$

	Draw the force diagram with friction acting down the slope against the direction of motion.
	Resolve perpendicular first to find R.
	Resolve parallel, assigning up the slope as positive, to find a.
	Use an equation of constant acceleration to find the distance.
	Draw a new force diagram because the situation has changed and friction now acts up the slope to prevent motion down the slope. Friction is limiting because we are assuming there will be motion down the slope.
	Resolve perpendicular to find the new value for R, which in this case is the same as the old value.
	Resolve parallel, assigning down the slope as positive, to find a.
	The acceleration down the slope should come out as positive to be consistent with the assumption that there is motion down the slope.
	This could also be done by calling the friction F and finding the size of F required to prevent motion and showing $F > \mu R$.
	Use an equation of constant acceleration to find the speed.

EXERCISE 4C

1. At the end of a downhill run, a skier of mass 80 kg slides up a rough slope at an angle of 10° to the horizontal, to slow down. He arrives at the upward slope with an initial speed of $12\,\text{m s}^{-1}$. The coefficient of friction between the skier and the slope is 0.4. Find how far up the slope he comes to rest, and show that he remains at rest there without falling back down the slope.

2 A ball of mass 3 kg rolls with initial speed 8 m s^{-1} up a rough slope at an angle of 15° to the horizontal. The coefficient of friction between the ball and the slope is 0.6.

   a By modelling the ball as a particle, find how long it takes for the ball to come to rest and show that the ball remains at rest there.

   b Why is this model different from reality?

3 A book of mass 3 kg is at rest on a rough slope at an angle of 15° to the horizontal. It takes a force of 20 N parallel to the slope to break equilibrium and drag it up the slope.

   a Find the coefficient of friction between the slope and the book.

   b Find the acceleration of the book down the slope if the 20 N force is applied down the slope.

4 A box of mass 12 kg is at rest on a rough slope at an angle of 18° to the horizontal. The coefficient of friction between the slope and the box is 0.4.

   a Find the force it takes parallel to the upwards slope to break equilibrium and drag the box up the slope.

   b If the force were applied down the slope and parallel to it, find the acceleration.

5 A car of mass 1250 kg is at rest on a rough slope at an angle of 35° to the horizontal. It takes a force of 13 000 N to move it up the slope. Show that without any force the car would slide down the slope, and find the minimum force to prevent it moving down.

6 A bin of mass 10 kg is at rest on a rough slope at an angle of 32° to the horizontal. It is held on the point of moving up the slope by a force of 90 N parallel to the slope. Show that when the force is removed the bin would slide down the slope, and find its acceleration.

7 A trolley of mass 5 kg is rolling up a rough slope, which is at an angle of 25° to the horizontal. The coefficient of friction between the trolley and the slope is 0.4. It passes a point A with speed 12 m s^{-1}. Find its speed when it passes A on its way back down the slope.

8 A ball of mass 1.5 kg is sliding up a slope, which is at 30° to the horizontal. The coefficient of friction between the ball and the slope is 0.45. It passes a point A at 10 m s^{-1}. By modelling the ball as a particle, find the time taken to return to A.

9 A pinball game involves hitting a ball up a slope whenever it reaches the bottom of the slope. The pinball has mass 0.2 kg and rolls down a rough slope of length 1.2 m at angle 12° to the horizontal and with coefficient of friction 0.1. The ball starts at the top of the slope at rest. When it reaches the bottom of the slope it is hit back up and its speed is increased by 50%.

   a Find the maximum height up the slope the pinball reaches after it has been hit back up the slope.

   b What assumptions have been made to answer the question?

10 A wooden block of mass 3.5 kg is sliding up a rough slope and passes a point A with speed 20 m s^{-1}. The slope is at 29° to the horizontal. The block comes to rest 25 m up the slope. Find its speed as it passes point A on the way down.

11 A boy drags a sledge of mass 4 kg from rest down a rough slope at an angle of 18° to the horizontal. He pulls it with a force of 8 N for 3 s by a rope that is angled at 10° above the parallel down the slope. After 3 s the rope becomes detached from the sledge. The coefficient of friction between the slope and the sledge is 0.4. Find the total distance the sledge has moved down the slope from when the boy started dragging it until it comes to rest.

12 A particle slides up a slope at angle 34° to the horizontal. It passes a point P on the way up the slope with speed $3\,\mathrm{m\,s^{-1}}$ and passes it on the way down the slope with speed $2\,\mathrm{m\,s^{-1}}$. Find the coefficient of friction between the particle and the slope.

P 13 A particle slides up a slope at angle θ to the horizontal with coefficient of friction μ. It passes a point A on the way up the slope at speed $u\,\mathrm{m\,s^{-1}}$ and passes it on the way down the slope with speed $v\,\mathrm{m\,s^{-1}}$. Prove that:

$$v^2 = u^2 \left(\frac{\sin\theta - \mu\cos\theta}{\sin\theta + \mu\cos\theta} \right)$$

so v is independent of the mass of the particle and the value of g. Deduce also that the speed on the way down is always smaller than the speed on the way up.

14 A toy car of mass 80 g rolls from rest 80 cm down a rough slope at an angle of 16° to the horizontal. When it hits a rubber barrier at the bottom of the slope it bounces back up the slope with its speed halved, and reaches a height of 10 cm. Find the coefficient of friction between the car and the slope.

4.4 Angle of friction

The concept of the **angle of friction** is not required by the syllabus. However, an understanding of the angle of friction can make some problems on the syllabus easier to solve and can help give alternative methods to solve problems on topics beyond the syllabus.

If a box is being pulled horizontally by a rope with tension T and is on the point of slipping, the force diagram would look like the first diagram. This diagram has four forces, but we can draw a simpler diagram with only three forces if we combine the normal contact force and friction into a single contact force, C, as shown in the second diagram.

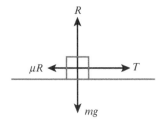 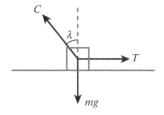

The angle of friction λ is the angle between the normal contact force and the total contact force when friction is limiting.

By drawing the components of the total contact force in a right-angled triangle it can be seen that $\tan\lambda = \dfrac{\mu R}{R} = \mu$.

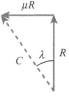

KEY POINT 4.9

The angle of friction is related to the coefficient of friction by $\lambda = \tan^{-1}\mu$.

By considering the total contact force as a single force rather than two forces (the normal contact force and friction), problems like the previous one with four forces can be reduced to problems with three forces. This means that you can use methods involving the triangle of forces or Lami's theorem.

In the simplest case of an object on a slope in limiting equilibrium under gravity, a problem with three forces becomes a problem with two forces.

REWIND

Look back to Chapter 3, Section 3.3, if you need a reminder of the triangle of forces and Lami's theorem.

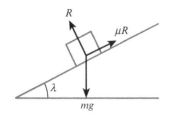
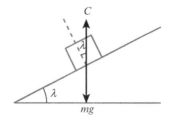

It is now easy to determine the total contact force. The contact force and the weight must be equal in magnitude and act in opposite directions, so the contact force will be vertical.

The angle between the normal contact force and the total contact force is the angle between the normal contact force and the vertical. This is also equal to the angle between the slope and the horizontal. If the object is in limiting equilibrium, then $\tan \lambda = \mu$ and we have the coefficient of friction.

KEY POINT 4.10

The angle of friction is the steepest slope on which an object can remain at rest without slipping under gravity.

EXPLORE 4.4

The coefficient of friction can be found by experiment using the angle of friction. Two people will be required to find it safely.

To find the coefficient of friction between an object and a table, place the object on the table. Then gradually lift one side of the table so the surface is at an angle to the horizontal. At the point where the object starts to slip down the table, remove the object from the table and measure the angle between the table and the horizontal. Use the equation $\tan \lambda = \mu$ to find the coefficient of friction.

WORKED EXAMPLE 4.8

A man tries to drag a suitcase of mass 18 kg along a rough horizontal surface. He drags it with a rope at an angle of 20° above the horizontal. The coefficient of friction between the ground and the suitcase is 0.4. The suitcase is in limiting equilibrium. Find the tension in the rope.

Answer

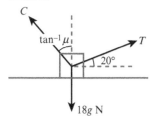

Mark the contact force as a single force so the problem now has only three forces.

There is limiting equilibrium, so the angle between C and the normal is $\tan^{-1} \mu$.

$$\frac{T}{\sin(180 - \tan^{-1}\mu)} = \frac{18g}{\sin(70 + \tan^{-1}\mu)}$$

We can now apply Lami's theorem.

$$T = 66.9 \text{ N}$$

WORKED EXAMPLE 4.9

a A 2 kg brick is at rest on a plank. The plank is lifted at one end to make an angle of 20° with the horizontal and the brick remains stationary on the plank. Find the total contact force between the brick and the plank.

b The plank is lifted further to an angle of 25° and the brick is on the point of slipping down the slope. Find the coefficient of friction between the plank and the brick.

c The plank is lifted further to an angle of 35° and the brick is held in place by a force at an angle of 15° above the angle of the upwards slope. Find the size of the force.

Answer

a

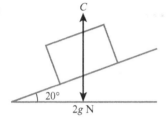

The only two forces are the total contact force and the weight, so they must cancel each other out by being equal in magnitude but opposite in direction.

$C = 2g$
$= 20$ N

Resolving vertically gives C immediately.

b

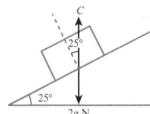

As before, the total contact force must be vertically upwards, so the angle between the normal and the total contact force is 25°.

$\mu = \tan 25$
$= 0.466$

Because the brick is now in limiting equilibrium, the coefficient of friction is found from the tan of the angle between the contact force and the perpendicular.

c

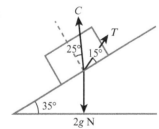

Using the total contact force, the force diagram now has only three forces in it and the triangle of forces can be used.

Since there is limiting friction, the angle of friction between C and the perpendicular to the slope will be the same as in the previous part, which was also limiting.

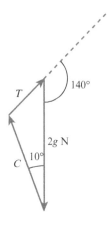

Angles in the triangle can often be found more easily by extending the sides if required or by comparing with vertical or horizontal lines on the original diagram.

$$\frac{T}{\sin 10} = \frac{2g}{\sin 130}$$

$$T = \frac{2g \sin 10}{\sin 130}$$

$$= 4.53 \text{ N}$$

Use the sine rule to find T.

EXERCISE 4D

1. A boy is trying to drag a box along a rough horizontal surface by pulling horizontally. The box has mass 12 kg. The coefficient of friction between the box and the surface is 0.4. The box is on the point of slipping. Find the size of the force exerted by the boy.

2. A girl tries to drag the box of mass 12 kg along the rough horizontal surface with coefficient of friction 0.4. She exerts a force at an angle of 10° above the horizontal and the box is on the point of slipping. Find the size of the force exerted by the girl.

3. A builder tries dragging a sack of sand of mass 25 kg along a rough horizontal surface with coefficient of friction 0.5. He pulls at 12° above the horizontal and the sack is on the point of slipping. Find the size of the total contact force.

4. A bench has mass 17 kg and is at rest on horizontal ground. A woman tries to move it by pulling it with a force of 80 N at 15° above the horizontal and the bench is on the point of slipping.

    a Find the angle of friction.

    b Hence, find the coefficient of friction between the bench and the ground.

5. A trailer has mass 120 kg. A winch pulls the trailer with a force of 500 N at an angle θ above the horizontal. The trailer is in limiting equilibrium on horizontal ground with coefficient of friction 0.45. Find θ.

6. A metal block of mass 20 kg is on a rough slope at an angle of 12° to the horizontal. The coefficient of friction between the book and the slope is 0.4. A boy is trying to move the block up the slope by pushing parallel to the slope. He increases the force until equilibrium breaks. Find the maximum size of the force the boy pushes with before the block slips.

7. A girl drags a sledge up a rough slope, which has an angle of 10° to the horizontal. The sledge has mass 8 kg and the coefficient of friction between the slope and the sledge is 0.3. She pulls the sledge with a rope at an angle of 12° to the slope and increases the tension until equilibrium is broken. Find the tension in the rope when this happens.

8 A car is towed down a rough slope, which is at an angle of 5° to the horizontal. The coefficient of friction between the car and the slope is 0.35. The car is towed using a rope at an angle of 13° to the slope. Equilibrium is broken when the tension in the rope is 4000 N. Find the mass of the car.

9 A box of mass 12 kg is at rest on a rough horizontal surface with coefficient of friction 0.6. A force is exerted on it at an angle θ above the horizontal so that the force required to break equilibrium is minimised. Show that θ is the angle of friction and find the size of the force required to break equilibrium.

10 A box has mass 40 kg and is on a rough slope with coefficient of friction 0.3. It is pulled up the slope by a force of 300 N at 10° above the slope and is in limiting equilibrium. Find the angle that the slope makes with the horizontal.

11 A ring of mass m kg is at rest on a fixed rough horizontal wire with coefficient of friction μ. It is attached to a string that is at an angle of α above the horizontal. Show that when $T < \dfrac{mg \sin \alpha}{\cos(\alpha - \theta)}$ and $\theta = \tan^{-1} \mu$ the ring will be in equilibrium.

Show further that if $\alpha + \theta \leq 90°$ and $T > \dfrac{mg \sin \alpha}{\cos(\alpha - \theta)}$ the ring will always move, but if $\alpha + \theta > 90°$ and $T > \dfrac{mg \sin \theta}{\sin(\alpha + \theta - 90)}$ the ring will remain in equilibrium.

12 A particle of weight W is at rest on a rough slope, which makes an angle of α to the horizontal. The coefficient of friction between the particle and the slope is μ. Assuming $\theta + \alpha < 90°$, where $\theta = \tan^{-1} \mu$, show that the minimum force F required to break equilibrium and make the particle slide up the slope is $F = W \sin(\theta + \alpha)$ and that F makes an angle θ to the slope above the particle.

Show further that in the case where $\alpha < \theta$, the minimum force F required to break equilibrium and make the particle slide down the slope is $F = W \sin(\theta - \alpha)$ and that F makes an angle θ to the slope below the particle.

Checklist of learning and understanding

- Friction can take any value up to the limiting value, which depends on the normal contact force, R, and the coefficient of friction, μ.
- $F \leq \mu R$
- If there is motion, or the object is on the point of slipping or in limiting equilibrium, friction will take the maximum possible value.
- The total contact force is the combination of the normal contact force and the friction.
- If a situation becomes different because a force changes or the direction of motion switches, the normal contact force may be affected, so the friction may change. It is best to draw a new diagram every time a different situation arises.

END-OF-CHAPTER REVIEW EXERCISE 4

1. A horizontal force, T, acts on a particle of mass 12 kg, which is on a rough horizontal plane. Given that the particle is on the point of slipping and that the coefficient of friction is 0.35, find the size of T.

2. A particle of mass 15 kg is on a slope at an angle 25° to the horizontal. The coefficient of friction between the particle and the slope is 0.3. A force, P, acts up the slope along the line of greatest slope. Find the set of values for P for the particle to be in equilibrium.

3. A bowler rolls a ten-pin bowling ball of mass 4 kg along a horizontal lane. The ball is released with a speed of $9 \,\text{m s}^{-1}$. The coefficient of friction between the ball and the lane is 0.04. The first pin is 18.5 m away. Find the speed at which the ball hits the pin.

4. A brick of mass 4.3 kg is being pushed up a slope by a force of 40 N parallel to the slope. The slope is at 13° to the horizontal and the coefficient of friction between the brick and the slope is 0.55. Find the acceleration of the brick.

5. A boat of mass 5 tonnes is being launched from rest into the sea by sliding it down a ramp. The ramp is at 5° to the horizontal and is lubricated so the coefficient of friction is only 0.08. The ramp is 40 m long before the boat enters the sea. Find the speed with which the boat enters the sea.

6. A bag of mass 49 kg is on rough horizontal ground with coefficient of friction 0.3. A force T acts at θ above the horizontal, where $\sin\theta = \dfrac{3}{5}$ and the bag is in limiting equilibrium. Show that $R = \dfrac{8T}{3}$, where R is the normal contact force, and find another equation relating R and T. Hence, find R and T.

7. A book of mass 1.3 kg is on a plank of wood, which is held at an angle of 16° to the horizontal. The coefficient of friction between the book and the plank is 0.45.
   a. Show that the book remains at rest and find the size of the frictional force.
   b. The book is held stationary while the plank is raised to make an angle of 27° with the horizontal. Show that when the book is released it accelerates down the slope, and find the size of the acceleration.

8. Two boys are arguing over who gets to play with a toy. The toy has mass 3 kg and is at rest on rough horizontal ground with a coefficient of friction of 0.3. The older boy pulls with a force of 26 N at an angle of 39° above the horizontal. The younger boy pulls in the opposite direction with a force of 24 N at an angle of 9° above the horizontal. Determine whether the toy moves. If it accelerates, find the size of the acceleration and direction. If not, find the size of the friction.

9. A mass of 6 kg is on a slope at an angle of 14° to the horizontal. The coefficient of friction between the slope and the mass is 0.4. There is a force of 5 N acting down the slope and parallel to it.
   a. Show that the force is not great enough to overcome friction, and find the magnitude of the total contact force between the mass and the slope.
   b. When the force of 5 N is removed, find the total contact force and the angle it makes with the slope.

10. A box of mass 9 kg rests on a slope, which is at an angle of 34° to the horizontal. It is held in place by a horizontal force of 20 N.
    a. By considering the total contact force as a single force, or otherwise, find the size of the total contact force.
    b. Given that friction is limiting, find the coefficient of friction between the box and the slope.

11 A particle of mass 6 kg is on a slope at an angle of 20° to the horizontal. The coefficient of friction between the particle and the slope is 0.1. The particle is 5 m from the bottom of the slope. It is projected up the slope with speed 4 m s^{-1}.

   a Find the distance travelled up the slope from the starting point until the particle comes to rest.

   b Find the time until the particle reaches the bottom of the slope.

12 A particle of mass 8 kg is at rest on a slope at angle 15° to the horizontal. The coefficient of friction between the particle and the slope is 0.05. The particle is pulled up the slope by a rope with tension 30 N at an angle of 20° above the line of the slope.

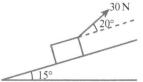

   a Find the acceleration of the particle.

   After travelling 10 m the string is cut and there is no tension.

   b Find the speed of the particle when the string is cut.

   The particle slows down until coming to rest.

   c Find how far the particle has travelled in total when it reaches its highest point on the slope.

   d Find the total time until it reaches that point.

P 13 A particle of mass m is on rough, horizontal ground with coefficient of friction μ_1. It is initially moving at speed u m s^{-1}. After a distance x m the surface changes to another surface with coefficient of friction μ_2. The particle comes to rest, having travelled a distance of y m on this surface. Show that $\mu_2 = \dfrac{u^2 - 2\mu_1 gx}{2gy}$.

P 14 A mass of m is at rest on a plank of wood on level ground with coefficient of friction μ_1. One end of the plank is lifted until the mass starts to slip. The angle at which this happens is α.

   a Show that $\mu_1 = \tan \alpha$.

   The angle of the plank is then raised to an angle β and the mass is held in place. The mass is then released and travels a distance x down the slope. At the end of the slope the particle slides along the level ground, slowing down under friction where the coefficient of friction is μ_2, until coming to rest at a distance y from the bottom of the slope. You may assume the mass starts sliding along the floor at the same speed as it has when it reaches the end of the slope.

   b Show that $\mu_2 = \dfrac{x(\sin \beta - \tan \alpha \cos \beta)}{y}$.

15 A particle moves up a line of greatest slope of a rough plane inclined at an angle α to the horizontal, where $\sin \alpha = 0.28$. The coefficient of friction between the particle and the plane is $\dfrac{1}{3}$.

   i Show that the acceleration of the particle is -6 m s^{-2}. [3]

   ii Given that the particle's initial speed is 5.4 m s^{-1}, find the distance that the particle travels up the plane. [2]

   Cambridge International AS & A Level Mathematics 9709 Paper 43 Q1 November 2013

16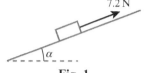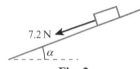

   Fig. 1   Fig. 2

A block of weight 7.5 N is at rest on a plane which is inclined to the horizontal at angle α, where $\tan\alpha = \dfrac{7}{24}$.

The coefficient of friction between the block and the plane is μ. A force of magnitude 7.2 N acting parallel to a line of greatest slope is applied to the block. When the force acts up the plane (see Fig. 1) the block remains at rest.

i   Show that $\mu \geqslant \dfrac{17}{24}$. [4]

When the force acts down the plane (see Fig. 2) the block slides downwards.

ii   Show that $\mu < \dfrac{31}{24}$. [2]

Cambridge International AS & A Level Mathematics 9709 Paper 41 Q3 November 2014

17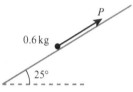

The diagram shows a particle of mass 0.6 kg on a plane inclined at 25° to the horizontal. The particle is acted on by a force of magnitude P N directed up the plane parallel to a line of greatest slope. The coefficient of friction between the particle and the plane is 0.36. Given that the particle is in equilibrium, find the set of possible values of P. [9]

Cambridge International AS & A Level Mathematics 9709 Paper 43 Q6 November 2012

Chapter 5
Connected particles

In this chapter you will learn how to:

- use Newton's third law for objects that are in contact
- calculate the motion or equilibrium of objects connected by rods
- calculate the motion or equilibrium of objects connected by strings
- calculate the motion or equilibrium of objects that are moving in elevators.

PREREQUISITE KNOWLEDGE		
Where it comes from	What you should be able to do	Check your skills

Chapter 1	Use the *suvat* equations for motion with constant acceleration.	1 A ball is thrown vertically upwards with initial speed $5\,\mathrm{m\,s^{-1}}$.
		  a How high does it rise?
		  b How long does it take to reach the maximum height?

Chapter 2	Use Newton's second law, know that weight = mg and know about normal contact forces.	2 A box of mass 20 kg is at rest on a smooth surface.
		  a Work out the normal contact force.
		  b What assumptions have you made in answering part a?

Chapter 3	Resolve forces in equilibrium and deal with non-equilibrium problems.	3 A box is sitting on a slope that makes an angle 30° with the horizontal. The box has weight 4 N.
		  a Work out the frictional force that is preventing the box from slipping down the slope.
		  The angle that the slope makes with the horizontal is increased to θ. The box starts from rest and slides with constant acceleration $2.5\,\mathrm{m\,s^{-2}}$ down the slope.
		  b Work out the frictional force in this situation.

How is the motion of an object affected by it being attached to something else?

When a car tows a trailer, the motion of the car is altered by the trailer. Mostly this is due to the extra weight of the trailer, although there may be additional resistance forces on the trailer. Would the motion of the car be the same if the trailer and its contents could be put inside the car?

In this chapter you will study the forces acting on different types of connected objects and look at how these forces affect or prevent motion. In particular, you will consider objects connected by rigid rods (such as a car towing a trailer), objects connected by strings (such as masses hanging on the ends of a rope that passes over a pulley) and objects in moving lifts (elevators). You will not be considering objects such as planets that affect each other 'remotely' using gravitational attraction.

You will use Newton's second law to calculate the acceleration of moving systems and Newton's third law to calculate normal contact forces (normal reaction forces).

5.1 Newton's third law

Newton's third law states that for every action there is an equal and opposite reaction. This means that in every interaction there is a pair of forces that have the same magnitude but act in opposing directions.

For example, when a boy uses a rope to pull a box, the force with which the box pulls on the boy is equal and opposite to the force with which the boy pulls the box. In both cases, the force is the tension in the rope.

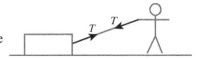

When two objects are in contact, each pushes on the other with an equal and opposite normal contact force.

A box resting on the floor pushes down on the floor with a vertical contact force and the floor pushes up on the box with an equal and opposite contact force. If no other forces act, these contact forces are each equal to the weight of the box.

A box resting on a slope pushes into the slope with a contact force and the slope pushes back with an equal and opposite contact force. When you draw a force diagram, you usually show only the forces acting *on* the box, so here you show the normal contact force from the slope on the box.

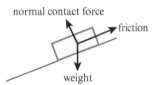

In Chapters 3 and 4, you studied the forces acting on an object and the motion or equilibrium of that object. You now do the same thing but for systems made up of connected objects.

5.2 Objects connected by rods

A **rod** is anything that can be modelled as a rigid connector with no mass. Examples of objects connected by rods include a car towing a caravan, a truck pulling a trailer, and a train made up of an engine pulling some carriages. In each of these situations you will only consider motion in a straight line; that is, in one dimension.

You can analyse the forces and the motion in these systems using Newton's second law.

KEY POINT 5.1

In a connected system, you can apply Newton's second law to the entire system or to the individual components of the system.

When you consider the individual components in a system of two objects connected by a rod such as a tow-bar, you need to include a **tension** force in the connecting rod or tow-bar. In some situations this tension may turn out to be negative. This means that the rod is under compression and the force is a **thrust**.

WORKED EXAMPLE 5.1

A car towing a trailer travels along a horizontal straight road. The car has mass 1500 kg and the trailer has mass 500 kg. The resistance to motion is 80 N on the car and 20 N on the trailer. The driving force produced by the engine of the car is 360 N. Find the tension in the tow-bar.

Answer

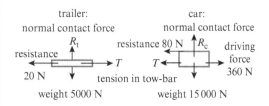

Newton's second law for the system:

$360 - 100 = 2000a$ so $a = 0.13\,\text{m s}^{-2}$

Newton's second law for the trailer and car separately:

$T - 20 = 500a$ and $360 - 80 - T = 1500a$

Hence, $T = 85\,\text{N}$.

> It is always a good idea to start with a diagram showing the forces.
>
> You can treat the system as a single entity to find the acceleration. This is because the internal tensions and thrusts cancel.
>
> There is no motion vertically, so the vertical components cancel out.
>
> The components must be treated separately when the internal tensions or thrusts are required.
>
> Draw separate diagrams to show the forces on the trailer and the forces on the car.
>
> Note that the force pulling the trailer forwards is the tension in the tow-bar.
>
> The car and trailer have the same acceleration.
>
> Either eliminate a or substitute $a = 0.13$.

WORKED EXAMPLE 5.2

A car towing a trailer travels along a horizontal straight road. The car has mass 1500 kg and the trailer has mass 500 kg. The resistance to motion is 80 N on the car and 20 N on the trailer. The driver applies the brakes, so the driving force is replaced by a braking force of 100 N opposing the forward motion.

a Find the force in the tow-bar.

The car then descends a hill at 3° to the horizontal. The resistances and braking force are unchanged.

b Find the new force in the tow-bar.

Answer

a

trailer:
normal reaction R_t

20 N ← → T
weight 5000 N

car:
normal reaction R_c
80 N
T ← 100 N
weight 15 000 N

> Draw the force in the tow-bar as a tension unless you know that it is a compression.

Newton's second law for the system:

$-100 - 80 - 20 = 2000a$ so $a = -0.1\,\text{m s}^{-2}$

> The resultant horizontal forces on the system are the braking force and the resistances.

Newton's second law for the trailer and car separately:

$T - 20 = 500a$ and $-100 - 80 - T = 1500a$

> The car and trailer have the same acceleration.

Hence, $T = -30$ N.

The force in the tow-bar is a thrust of 30 N.

b

Either eliminate a or substitute $a = -0.1$.

Draw a new force diagram for the new situation. The forces labelled R_t, R_c and T will not necessarily have the same values as in part **a**.

Newton's second law for the system (parallel to the slope):
$15\,000 \sin 3 + 5000 \sin 3 - 100 - 80 - 20 = 2000a$
so $a = 0.423\,\text{m s}^{-2}$.

Newton's second law for the trailer and car separately:
$T + 5000 \sin 3 - 20 = 500a$
and $15\,000 \sin 3 - 100 - 80 - T = 1500a$

Hence, $T = -30$ N.

In this case, the force in the tow-bar is still a thrust of 30 N.

The acceleration is parallel to the slope (down the slope). The angle between the vertical and the normal to the slope is 3°. The weight of the car has a component $15\,000 \sin 3$ down the slope. The component normal to the slope ($15\,000 \cos 3$) is balanced by R_c. Similarly for the weight of the trailer.

WORKED EXAMPLE 5.3

A model train consists of an engine (locomotive) coupled to a chain of four trucks. The coupling between the engine and the first truck and each coupling between trucks are modelled as rigid rods. The train is moving on a straight horizontal track. The engine has mass 0.8 kg and each truck has mass 0.2 kg when empty. The resistance to motion is 0.06 N on the engine and 0.01 N on each truck. The driving force produced by the engine is 3 N.

   a A mass of 0.1 kg is placed in each truck. Find the tension in each coupling.

   b Find the tension in each coupling if, instead, the 0.4 kg is all placed in the last truck.

Answer

a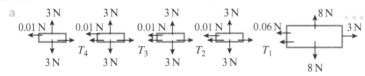

Draw a diagram to show the forces acting on the engine and each truck.

Newton's second law for the system:
$3 - (0.06 + 0.01 + 0.01 + 0.01 + 0.01) = (0.8 + 4 \times 0.1 + 4 \times 0.2)a$
$3 - 0.1 = 2a$
$a = 1.45\,\text{m s}^{-2}$

The resultant horizontal forces on the system are the driving force and the resistances.

Newton's second law for the engine and each truck separately:
$3 - 0.06 - T_1 = 0.8a$ $T_1 = 1.78$ N
$T_1 - 0.01 - T_2 = 0.3a$ $T_2 = 1.335$ N
$T_2 - 0.01 - T_3 = 0.3a$ $T_3 = 0.89$ N
$T_3 - 0.01 - T_4 = 0.3a$ $T_4 = 0.445$ N
and $T_4 - 0.01 = 0.3a$

The acceleration is the same for the engine and for each truck.

There are five equations and four unknowns. The 'spare' equation can be used to check the values.

b

6 N	2 N	2 N	2 N		8 N	
0.01 N ↑	0.01 N ↑	0.01 N ↑	0.01 N ↑	0.06 N	↑	3 N
← □ → T_4	← □ → T_3	← □ → T_2	← □ → T_1	←	□	→
↓ 6 N	↓ 2 N	↓ 2 N	↓ 2 N		↓ 8 N	

Draw a new force diagram.

Newton's second law for the system:

$3 - 0.1 = 2a$

$a = 1.45\,\text{m s}^{-2}$

The resultant horizontal forces on the system are the driving force and the resistances.

Newton's second law for the engine and each truck separately:

$3 - 0.06 - T_1 = 0.8a$ $\quad T_1 = 1.78\,\text{N}$
$T_1 - 0.01 - T_2 = 0.2a$ $\quad T_2 = 1.48\,\text{N}$
$T_2 - 0.01 - T_3 = 0.2a$ $\quad T_3 = 1.18\,\text{N}$
$T_3 - 0.01 - T_4 = 0.2a$ $\quad T_4 = 0.88\,\text{N}$
and $T_4 - 0.01 = 0.6a$

The acceleration is the same for the engine and for each truck.

EXERCISE 5A

1 A tractor is connected to a trailer by a rigid, light bar. The tractor has mass 10 000 kg and the trailer has mass 2000 kg. The tractor and trailer are moving along a straight horizontal road. The tractor engine produces a driving force of 400 N. The resistance on the tractor is 40 N and the resistance on the trailer can be ignored. Find the tension in the bar.

2 A car of mass 1200 kg tows a trailer of mass 300 kg. The car and trailer travel along a straight horizontal section of road. The engine of the car produces a driving force of 400 N. The car experiences a resistance of 150 N and the trailer experiences a resistance of 100 N.

    a Find the acceleration of the car and trailer.

    b Find the tension in the tow-bar.

M 3 A box of weight 250 N is pulled across a smooth horizontal floor, using a horizontal rope. The tension in the rope is 20 N.

    a Work out the acceleration of the box.

    b What modelling assumptions have been made?

A box of weight 100 N is connected to a second box of weight 150 N, using a connecting rod. The 150 N box is pulled across a smooth horizontal floor, using a horizontal rope. The tension in this rope is 20 N and air resistance can be ignored.

    c Work out the tension in the connecting rod between the two boxes.

    d Work out the tension in the connecting rod if the rope is attached to the 100 N box instead, but otherwise the situation is unchanged.

4 A truck of mass 2000 kg tows a trailer of mass 800 kg. The engine of the truck produces a driving force of 300 N. A resistance of 120 N acts on the truck and a resistance of 40 N acts on the trailer. The truck and trailer are moving along a straight horizontal road and initially the trailer is empty.

   a Find the tension in the tow-bar when the trailer is empty.

   A load of mass 1200 kg is then added to the trailer, which increases the resistance on the trailer to 80 N. The forces on the truck are unchanged. The truck and trailer return along the same straight horizontal road.

   b Find the tension in the tow-bar when the trailer carries this load.

5 A car of mass 2000 kg pulls a caravan of mass 1200 kg along a straight horizontal road. The resistance on the car is 20 N and the resistance on the caravan is 80 N. The maximum possible driving force from the car's engine is 1900 N. The tow-bar will break if the tension exceeds 680 N.

   a Find the maximum possible driving force before the tow-bar breaks.

   b Find the maximum possible acceleration.

6 A bucket hangs from a vertical rod. Another rod is attached to the bottom of the bucket and a second bucket hangs on the end of this rod. Each bucket is partially filled with water and they hang in equilibrium.

   a Work out the tension in each rod when:

   i each bucket of water has mass 12 kg

   ii the first bucket of water has mass 8 kg and the second has mass 16 kg.

   b What assumptions have been made?

7 A 15 kg mass hangs in equilibrium on a heavy chain. The chain is modelled as ten 1 kg masses joined by short rods of negligible mass. Including the connection at each end of the chain, this makes 11 short rods. Work out the tension in each of the 11 short rods.

8 A horizontal bar of mass 1 kg hangs from a pair of parallel vertical rods, of negligible mass, attached to either end of the bar. A third vertical rod is connected to the middle of the bar and a 4 kg mass hangs from this, below the rod. Work out the tension in each of the rods.

9 A train consists of an engine and five carriages. The engine has mass 100 000 kg and each carriage has mass 20 000 kg. The engine produces a driving force of 350 000 N. The resistance force on the engine is 10 000 N and the resistance on each carriage is 2000 N. The train moves in a straight line on a horizontal track. Find the tension in the coupling between the third carriage and the fourth carriage.

10 A car pulls a caravan up a hill. The slope of the hill makes an angle θ with the horizontal, where $\sin\theta = \dfrac{1}{20}$.

The car has mass 1900 kg and the caravan has mass 600 kg. The driving force from the engine of the car is 1200 N. The resistance on the car is 20 N and that on the caravan is 80 N. Find the force in the tow-bar and state whether it is a tension force or a thrust force.

11 A car tows a caravan down a hill. The slope of the hill makes an angle θ with the horizontal, where $\sin\theta = \dfrac{1}{20}$.

The car has mass 1900 kg and the caravan has mass 600 kg. The car is braking, so the driving force from the engine of the car is negative. This braking force is 250 N (a driving force of −250 N). The resistance on the car is 20 N and that on the caravan is 80 N. Find the force in the tow-bar and state whether it is a tension force or a thrust force.

Cambridge International AS & A Level Mathematics: Mechanics

  12 A car tows a caravan down a hill. The slope of the hill makes an angle θ with the horizontal, where $\sin\theta = 0.05$. The force from the car's engine is a braking force (a negative driving force). The car has mass 1800 kg and the caravan has mass 600 kg. The resistance on the car is 20 N and that on the caravan is 80 N. The force in the tow-bar is a thrust of 50 N. Show that the force from the car's engine is −420 N.

5.3 Objects connected by strings

There are three main differences between rods and **strings**:

- a string can change direction (for example, by passing over a smooth peg or pulley)
- a string can be in tension or be slack (that is, have no tension)
- the force in a string can never be a thrust.

We use the term 'string' to mean any rope, chain or cable. You will always assume that the string is light, so its weight can be ignored.

To keep things simple, you also assume that any strings are inextensible (they do not stretch).

KEY POINT 5.2

When a string passes over a **smooth pulley**, the magnitude of the tension is unchanged but the direction can change.

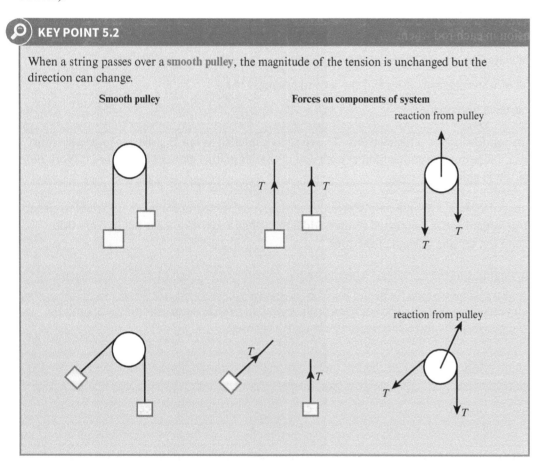

It is wrong to apply Newton's second law when the direction of travel changes. (Even if this gives the right answer, it is mathematically wrong to 'bend' a vector round a corner.)

When the acceleration of a system of connected objects is constant, you can use the equations for constant acceleration to calculate velocities, distance travelled or time taken.

DID YOU KNOW?

Archimedes invented many machines as 'amusements in geometry', some of which were used to defend Syracuse when it was attacked in 212 BCE by the Romans. These machines used levers and pulleys; for example, an 'iron hand' that could lift the Roman ships into the air and swing them to and fro until all the Roman soldiers were thrown out.

He also invented a compound pulley that brought him great fame when he used it to move a fully laden ship with a crew of many men 'as smoothly and evenly as if she had been at sea', by holding the head of the pulley in his hand and pulling on the cords.

WORKED EXAMPLE 5.4

A box of mass 3 kg is placed on a table. The coefficient of friction between the box and the table is 0.5. A string is attached to the box and passes over a smooth pulley at the edge of the table. The part of the string between the box and the pulley is horizontal. After passing over the pulley, the string hangs vertically, with the other end attached to a ball of mass 2 kg. The system is released from rest.

a Find the tension in the string.

The ball is initially 8 cm below the table top. The ball hits the ground after 1.2 s. The box has not reached the pulley at this time.

b Find the height of the table.

Answer

a

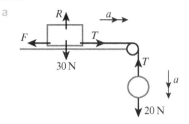

The forces acting on the box are:
- its weight (30 N)
- the normal reaction from the contact with the table (R)
- the tension in the string (T)
- the frictional resistance (F).

The forces acting on the ball are:
- its weight (20 N)
- the tension in the string (T).

The two tensions are numerically equal because the pulley is smooth.

The (horizontal) acceleration of the box is numerically equal to the (vertical) acceleration of the ball since otherwise the string would either snap or would go slack.

Resolving vertically for the box:

$R = 30$

The box does not move vertically so there is no resultant force vertically.

$F = \mu R$

$= 0.5 \times 30$

$= 15$

The box is moving so friction is at its limiting value.

Cambridge International AS & A Level Mathematics: Mechanics

Newton's second law for the box (horizontally, left to right):
$T - F = 3a$
$T - 15 = 3a$
Newton's second law for the ball (vertically downwards):
$20 - T = 2a$

> The box and the ball are moving in different directions, so we must treat the (horizontal) motion of the box and the (vertical) motion of the ball separately.

Hence, $2T - 30 = 60 - 3T$.

So $T = 18$.

The tension in the string is 18 N.

> Solve the equations simultaneously by eliminating a.

b $a = 1$

> Substitute $T = 18$ into $20 - T = 2a$.

The box and the ball each accelerate at $1\,\text{m s}^{-2}$.

For the ball:
$u = 0\,\text{m s}^{-2}$, $a = 1\,\text{m s}^{-2}$, $t = 1.2\,\text{s}$

$s = ut + \dfrac{1}{2}at^2$

> Use an equation for constant acceleration.

$= \dfrac{1}{2} \times 1 \times 1.2^2$

$= 0.72$

The ball falls 0.72 m. So the height of the table is $72 + 8 = 80\,\text{cm}$.

KEY POINT 5.3

You can apply Newton's second law to any part of the connected system in which all objects are moving with the *same acceleration* and in the *same direction*.

WORKED EXAMPLE 5.5

A crate of mass 500 kg rests on a slope and is attached to a rope that passes over a smooth pulley. The slope is inclined at 20° to the horizontal. The coefficient of friction between the slope and the crate is 0.1.

What happens to the crate when a force of 2000 N is applied to the vertical part of the rope?

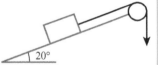

Answer

The pulley is smooth so the tension in the rope at the point of contact with the crate is also 2000 N.

If the crate slides up the slope:

> We are not told whether or not the crate moves up the slope, is stationary or moves down the slope.

Resolving perpendicular to the slope:
$R = 5000 \cos 20$
$ = 4700 \text{ N}$
The crate is sliding,
so $F = 0.1R$
$ = 470 \text{ N}$

Friction is limiting so $F = \mu R$.

Component of weight down slope $= 5000 \sin 20$
$ = 1710 \text{ N}$

Resultant force up slope $= 2000 - 470 - 1710$
$ = -180 \text{ N}$

This means that the acceleration up the slope would be negative, and since the crate is initially at rest we can conclude that it does not slide up the slope.

If the crate slides down the slope:

The only difference is that the friction now acts up the slope.

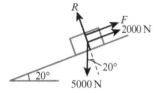

As before, $R = 5000 \cos 20$
$ = 4700 \text{ N}$
and $F = 0.1 \times R = 470 \text{ N}$
and component of weight down slope $= 1710 \text{ N}$

Resultant force up slope $= 2000 + 470 - 1710$
$ = 760 \text{ N}$

This means that the acceleration up the slope would be positive, and since the crate is initially at rest we can conclude that it does not slide down the slope.

Therefore, the frictional force is sufficient to allow the crate to rest on the slope without sliding.

An alternative approach is to calculate the frictional force needed for equilibrium and compare it to the limiting value of friction. The component of the weight down the slope is 1710 N and the force up the slope is 2000 N, so the forces are in equilibrium when the frictional force is $2000 - 1710 = 290 \text{ N}$ down the slope.

The limiting (maximum) value of F is $\mu R = 470 \text{ N}$.

This is greater than the frictional force needed for equilibrium, so, in this situation, friction is not limiting and is sufficient to prevent the crate from sliding.

Cambridge International AS & A Level Mathematics: Mechanics

> **MODELLING ASSUMPTIONS**
>
> In all the problems in this chapter you are making assumptions about the way that objects are connected.
>
> The mass of a real rod will affect the equation in Newton's second law, but if the mass is sufficiently small in comparison to the mass of the objects, the effect in the equation is negligible.
>
> If a string moves around a pulley, the mass of the string moving in the different directions on either side of the pulley will be constantly changing. This would make the problem much more complicated but, because strings tend to be much lighter than the objects they pull, you consider the mass to be negligible.
>
> You are assuming the string is inextensible, which means the objects on either side of the string accelerate at the same rate in the direction of the string. Strings generally do extend slightly under tension, but the extension is sufficiently small to ignore.
>
> You are assuming the pulley is smooth. This means that the tension in the string on either side is the same, and also that the string slides over the pulley. With friction, the pulley itself may rotate, and factors such as the mass of the pulley would affect the motion. This makes the problem much more complicated.

EXERCISE 5B

1. In each of the following diagrams, the blocks are at rest and are connected by light strings passing over smooth pulleys. Any hanging portion of a string is vertical and any other portion is parallel to the surface. Unless marked otherwise, the surfaces are rough and horizontal. In each case, find the magnitude of the tension in each string and the magnitude of any frictional force.

   a 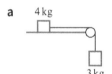   b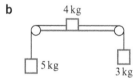

   c 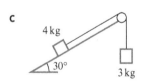   d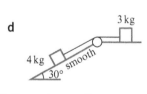

2. A bucket of mass 3 kg rests on scaffolding at the top of a building. The scaffolding is 22.5 m above the ground. The bucket is attached to a rope that passes over a smooth pulley. At the other end of the rope there is another bucket of mass 3 kg, which initially rests on the ground. The bucket at the top of the building is filled with 6 kg of bricks and is gently released. As this bucket descends, the other bucket rises.

   a Find how long it will take the descending bucket to reach the ground.

   b What modelling assumptions have been made?

3. A light inextensible string is fixed to a point on a ceiling. A box of mass 2 kg hangs from the string. Two light inextensible strings are attached to the box and hang vertically below the box. A particle of mass 0.4 kg hangs from the lower end of one string and a particle of mass 0.6 kg hangs from the lower end of the other string. Work out the tension in each string.

4 A block of mass 5 kg hangs from one end of a light inextensible string of length 2 m. The string passes over a small smooth pulley at the edge of a smooth horizontal table of height 70 cm. The other end of the string is connected to a block of mass 3 kg held at rest on the table. The portion of the string between the pulley and the 3 kg mass is horizontal and of length 1.5 m. The system is released from rest.

   a How long does it take for the 5 kg mass to reach the ground?

   b What is the speed of the 3 kg mass when the 5 kg mass hits the ground?

   The 3 kg mass continues to slide towards the pulley.

   c How long does it take, from when the system is released, for the 3 kg mass to reach the pulley?

5 A mass X, of 1 kg, hangs from one end of a light inextensible string. The string passes over a small, smooth, fixed pulley and a mass Y, of 0.6 kg, hangs from the other end of the string. Initially X is 0.4 m below the pulley and Y is 1.8 m below the pulley. The pulley is 3 m above the ground. The system is released from rest.

   a Find how long it takes from when the system is released until Y hits the pulley.

   When Y hits the pulley, the string breaks.

   b Find how long it takes from when the system is released until X hits the ground.

6 A block of mass 4 kg is held on a rough slope that is inclined at 30° to the horizontal. The coefficient of friction between the slope and the block is 0.2. A light inextensible string is attached to the block and runs parallel to the slope to pass over a small smooth pulley fixed at the top of the slope. The other end of the string hangs vertically with a block of mass 1 kg attached at the other end. The system is released from rest.

   a Work out the tension in the string.

   After 1.2 s the block of mass 4 kg reaches the bottom of the slope. The other block has not yet reached the pulley.

   b Work out a lower bound for the length of the string, giving your answer to 2 significant figures.

7 A particle of mass 0.3 kg hangs from one end of a light inextensible string. The string passes over a smooth pulley and a particle of mass 0.5 kg hangs from the other end. A second string is tied to the particle of mass 0.5 kg, and a particle of mass 0.2 kg hangs from this string, so that the particle of mass 0.2 kg hangs vertically below the particle of mass 0.5 kg. The system is released from rest. Find the tension in each string.

8 Two smooth pulleys are 8 m apart at the same horizontal level. A light inextensible rope passes over the pulleys and a box of mass 5 kg hangs at each end of the rope. A third box of mass m kg is attached to the midpoint of the rope and hangs between the pulleys so that all three boxes are at the same horizontal level.

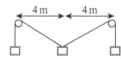

   The total length of the rope is 16 m. Find the value of m.

9 Two smooth pulleys are fixed at the same horizontal level. A light inextensible rope passes over the pulleys and a box of mass m kg hangs at each end of the rope. A third box of mass km kg is attached to the midpoint of the rope and hangs between the pulleys so that all three boxes are at the same horizontal level. The portions of the string that are not vertical make an angle of 30° with the horizontal. The system hangs in equilibrium. Find the value of k.

10 A mass of 2 kg is held on a rough horizontal table. The coefficient of friction between the table and the mass is 0.05. The 2 kg mass is attached by a light inextensible string to a mass of 3 kg and by a second light inextensible string to a mass of 4 kg. The strings pass over smooth pulleys at the edges of the table. The 3 kg mass hangs on one side of the table and the 4 kg mass hangs on the other side of the table. The system is released from rest. Find:

   a the acceleration of the masses
   b the tension in each string.

P **PS** **11** A wedge has two smooth sloping faces; one face makes an angle 30° with the horizontal and the other makes an angle 60° with the horizontal. A small smooth pulley is fixed at the apex of the wedge. A light inextensible string passes over the pulley and lies parallel to the faces of the wedge. At each end of the string there is a particle of mass 0.3 kg. The system is released from rest.

a Show that the tension in the string is $\dfrac{3(1+\sqrt{3})}{4}$ N.

b Work out the resultant horizontal force on each of the particles.

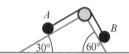

P **PS** **12** A box of mass 3 kg hangs from a light inextensible string, which passes over a smooth pulley fixed below a beam and then under a smooth cylinder of mass 4 kg that is free to move. The other end of the string is fixed to the beam.

The system is released from rest.

a Explain why the magnitude of the acceleration of the cylinder is half the magnitude of the acceleration of the box.

b Find the acceleration of the box, including its direction.

P **M** **13** A horizontal shelf of mass 1 kg hangs from four strings. A book of mass 0.2 kg sits on the shelf. Four strings are attached to the underside of the shelf and a second horizontal shelf of mass 1 kg hangs from these strings.

a What modelling assumptions can be made about the strings?

b Find the tension in each of the upper set of strings.

c Find the tension in each of the lower set of strings.

The book is moved to the lower shelf.

d How does this change the tensions in the strings?

EXPLORE 5.1

What would happen in the situation described in question 12 of Exercise 5B if the string passed over a second fixed pulley and under a second cylinder of mass 4 kg before being fixed to the beam? Investigate what happens if the masses are changed.

5.4 Objects in moving lifts (elevators)

When a person travels up or down in a lift, the floor of the lift acts as a connection between the person and the lift.

For the system (person in lift), the forces are the weights and the tension in the lift cable.

The forces on the person are their weight and the normal reaction from the floor of the lift.

The forces on the lift are the reaction from the person on the floor (which, by Newton's third law, is equal and opposite to the normal reaction from the floor on the person), the weight of the lift and the tension in the lift cable.

When you consider the lift and the person as a single object, the normal reaction forces cancel out.

When the lift is accelerating, the normal reaction from the person on the lift is not the same as the weight of the person.

Suppose that the tension in the cable is T, the normal reaction is R, the weight of the lift is W and the weight of the person is w, where these forces are all measured in newtons.

If the lift is accelerating upwards with acceleration $a\,\text{m s}^{-2}$, you can apply Newton's second law to the system:

$T - W - w = (M + m)a$

where m is the mass of the person and M is the mass of the lift, both in kg.

You can also apply Newton's second law to the person and the lift separately:

$R - w = ma$ and $T - R - W = Ma$

You can then calculate the acceleration of the lift, the tension in the cable or the normal reaction between the person and the floor.

If a is negative it could be because the lift is travelling upwards but slowing down or because it is travelling downwards and speeding up.

If a is positive it could be because the lift is travelling upwards and speeding up or because it is travelling downwards and slowing down.

WORKED EXAMPLE 5.6

A woman of mass 50 kg is travelling in a lift of mass 450 kg. The tension in the cable pulling the lift upwards is 5250 N. Calculate the acceleration of the lift.

Answer

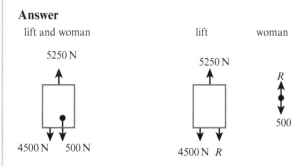

For the system:

$5250 - 4500 - 500 = (450 + 50)a$

$250 = 500a$

$a = 0.5\,\text{m s}^{-2}$

Apply Newton's second law to the whole system.

To find the acceleration you can work either with the entire system or with the individual components. In Worked example 5.6 we used the system and in Worked example 5.7 we show the same idea but using the individual components. To find the reaction forces you need to consider the individual components.

WORKED EXAMPLE 5.7

A man of mass 80 kg and a woman of mass 70 kg are travelling in a lift of mass 500 kg. The tension in the cable pulling the lift upwards is 6890 N. Calculate the acceleration of the lift and the reaction forces between the lift floor and each of the passengers.

Answer

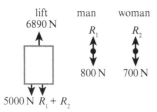

For the lift: $6890 - 5000 - R_1 - R_2 = 500a$
For the man: $R_1 - 800 = 80a$
For the woman: $R_2 - 700 = 70a$

$1890 - 800 - 700 = 650a$

$a = 0.6 \, \text{m s}^{-2}$

$R_1 - 800 = 80 \times 0.6 \quad R_1 = 848 \, \text{N}$ reaction from floor on man
$R_2 - 700 = 70 \times 0.6 \quad R_2 = 742 \, \text{N}$ reaction from floor on woman

· To find the reaction forces we need to use the individual components.

· Apply Newton's second law to the individual components.

· Add the equations to eliminate R_1 and R_2.

· Substitute $a = 0.6$ back into the previous equations to find R_1 and R_2.

EXERCISE 5C

1 A crate of mass 20 kg is put into a lift. The lift accelerates upwards at $0.3 \, \text{m s}^{-2}$. The tension in the lift cable is 5000 N.

   a Find the contact force between the lift floor and the crate.

   b Find the mass of the lift, giving your answer to the nearest kg.

2 A crate of mass 20 kg is put into a lift. The mass of the lift is 300 kg. Find the tension in the lift cable:

   a when the lift accelerates upwards at $0.3 \, \text{m s}^{-2}$

   b when the lift travels at constant speed

   c when the lift accelerates downwards at $0.3 \, \text{m s}^{-2}$.

PS 3 A man of mass 80 kg stands in a lift. The mass of the lift is 400 kg. The lift starts to travel downwards with an acceleration of $8 \, \text{m s}^{-2}$. The tension in the lift cable is kR, where R is the contact force between the man and the lift floor. Find the value of k.

4 A crate of mass 40 kg is put into a lift. The lift accelerates upwards at 0.4 m s^{-2}. The mass of the lift is 460 kg.
   a Find the tension in the lift cable.
   b Find the contact force between the lift floor and the crate.

5 A box of mass 20 kg sits on the floor of a lift. A second box of mass 10 kg sits on top of the first box and a third box of mass 5 kg sits on top of the second box. When the tension in the lift cable is 4620 N, the lift is accelerating upwards at 0.5 m s^{-2}.
   a Work out the mass of the lift.
   b Work out the reaction between the floor of the lift and the first box.
   c Work out the reaction between the first box and the second box.
   d Work out the reaction between the second box and the third box.

PS 6 The mass of a lift is 200 kg and the maximum tension in the lift cable is 2500 N.
   a Work out the maximum upwards acceleration of the lift when it is empty.

   The lift carries a load of 40 kg. The lift accelerates upwards with the maximum upwards acceleration possible.

   b Work out the contact force between the lift floor and the load.

7 A crate of mass m kg is put into a lift. The lift accelerates upwards at a m s^{-2}. The mass of the lift is M kg.
   a Find the tension in the lift cable.
   b Find the contact force between the lift floor and the crate.

P 8 A box of mass 5 kg sits in a lift. A second box of mass 3 kg sits on the first box. The lift accelerates upwards with acceleration 0.7 m s^{-2}.
   a Work out the contact force between the two boxes.

   The tension in the lift cable is unchanged but the boxes are swapped over, so the first box sits on the second box.

   b Show that this increases the contact force between the boxes.

PS 9 A passenger lift has mass 500 kg. The breaking tension of the cable is 12 000 N. The maximum acceleration of the lift is 0.75 m s^{-2}.
   a If the lift travels at its maximum acceleration, calculate the maximum mass of the passengers:
      i when the lift is accelerating upwards
      ii when the lift is accelerating downwards.
   b Taking the average mass of a person to be 75 kg, what is the maximum number of passengers that should be allowed to travel in the lift?

PS 10 The tension in a lift cable is 11 070 N. The lift is accelerating upwards at a m s^{-1}. A man of weight 800 N stands in the lift on a set of bathroom scales. The scales suggest that the weight of the man is 820 N. Assuming that the scales have negligible weight, find the weight of the lift.

11 Two masses, A of 2 kg and B of 3 kg, are connected by a light inextensible string that passes over a small, smooth, fixed pulley. Initially, the masses are held stationary and are then released.

   a Find the acceleration of each mass.

   The pulley is fixed to the wall of a lift. The lift is initially stationary. The lift has mass 400 kg and a man of mass 75 kg is travelling in the lift, along with the pulley system. There is nothing else in the lift. The lift starts to move upwards, from rest, by a tension in the lift cable of 4992 N.

   b Find the acceleration of the lift.

   c Find the acceleration of each of A and B, as viewed by a person standing stationary outside the lift.

Checklist of learning and understanding

- Newton's third law states that for every action there is an equal and opposite reaction. This means that in every interaction there is a pair of forces that have the same magnitude, but which act in opposing directions.

- When a string passes over a *smooth* pulley, the magnitude of the tension is unchanged but the direction can change.

- Newton's second law can be applied to a system of connected objects, either to the entire system or to individual components of the system, provided they move with the *same acceleration* and in the *same direction*.

Chapter 5: Connected particles

END-OF-CHAPTER REVIEW EXERCISE 5

1. A 10 m long slope makes an angle 30° with the horizontal. A box of mass 8 kg is pulled up the slope by a rope that is parallel to the slope. The coefficient of friction between the box and the slope is $\frac{1}{\sqrt{12}}$. At the top of the slope, the rope passes over a smooth pulley. A ball of mass 12 kg hangs from the other end of the rope. The ball is initially 2 m above the ground. The system is released from rest.

   a Find how long it will take for the box to travel 1.5 m up the slope. [3]

   b Find the tension in the rope. [2]

2. A car is travelling along a straight horizontal road. The car has mass 1800 kg and is towing a trailer of mass 600 kg. Resistance forces are 30 N on the car and 10 N on the trailer. Find the size and type of force in the tow-bar:

   a when the driving force from the engine is 400 N [3]

   b when the driving force from the engine is 20 N. [2]

3. A crate of mass 15 kg rests on a platform. The platform has mass 5 kg. It is lowered using a rope. The tension in the rope is 10 N.

   a Find the acceleration of the crate. [2]

   b Find the contact force between the platform and the crate. [3]

4. Particles P and Q, of masses 0.6 kg and 0.4 kg, respectively, are joined by a light inextensible string that passes over a small, smooth, fixed pulley. The particles are held at rest with the string taut and its straight parts vertical. Initially, both particles are 1.45 m above the ground and 0.25 m below the pulley. The system is released from rest.

   a Find the speed of P when Q reaches the pulley. [4]

   The string then breaks and P falls to the ground.

   b Find the time from when the system is released to when P hits the ground. [4]

5. Particles A and B, each of mass 0.5 kg, are attached to the ends of a light inextensible string. Particle A is held on a smooth slope inclined at 30° to the horizontal. The string passes over a small smooth pulley at the top of the slope and particle B hangs vertically below the pulley. Particle A is released and moves up the slope.

   a Find the acceleration of particle A up the slope. [3]

   Particle B hits the ground after 1.2 s. It then stays on the ground and particle A travels further up the slope. Particle A does not reach the pulley in the subsequent motion.

   b Find the distance travelled by particle A, from when it is released to when it comes to instantaneous rest. [5]

6. Particles A and B, of masses 0.5 kg and 2.5 kg, respectively, are attached to the ends of a light inextensible string. Particle A is held on a rough horizontal surface with coefficient of friction 0.2. The string passes over a small smooth pulley at the edge of the surface, at a distance 5 m from particle A. Particle B hangs vertically below the pulley. Particle A is released and particle B descends 1 m to reach the ground. When particle B reaches the ground, it stays there. Find the time taken from the start until particle A comes to instantaneous rest. [8]

7 A crate of mass 20 kg is pulled vertically upwards using a rope that passes over a first pulley, under a second pulley and over a third pulley. At the other end of the rope there is a ball of mass 30 kg. Each pulley has mass m kg. The first and third pulleys are fixed at the same horizontal level and are 4 m apart. The second pulley is an equal distance from the first and third pulleys and hangs at a distance 2 m below them. It is not fixed, but it does not move.

  a What modelling assumptions need to be made? Which of these assumptions is unlikely to affect the equilibrium of the second pulley? [3]

  b Find the value of m. [5]

8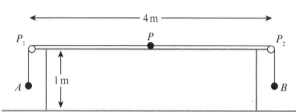

A light inextensible string of length 5.28 m has particles A and B, of masses 0.25 kg and 0.75 kg respectively, attached to its ends. Another particle, P, of mass 0.5 kg, is attached to the mid-point of the string. Two small smooth pulleys P_1 and P_2 are fixed at opposite ends of a rough horizontal table of length 4 m and height 1 m. The string passes over P_1 and P_2 with particle A held at rest vertically below P_1, the string taut and B hanging freely below P_2. Particle P is in contact with the table halfway between P_1 and P_2 (see diagram). The coefficient of friction between P and the table is 0.4. Particle A is released and the system starts to move with constant acceleration of magnitude a m s^{-2}. The tension in the part AP of the string is T_A N and the tension in the part PB of the string is T_B N.

  i   Find T_A and T_B in terms of a. [3]

  ii Show, by considering the motion of P that $a = 2$. [3]

  iii Find the speed of the particles immediately before B reaches the floor. [2]

  iv Find the deceleration of P immediately after B reaches the floor. [2]

Cambridge International AS & A Level Mathematics 9709 Paper 42 Q7 June 2014

9 Particles A and B, of masses 0.5 kg and 2.5 kg, respectively, are attached to the ends of a light inextensible string. Particle A is held on a rough slope. The slope is inclined at 30° to the horizontal and the coefficient of friction between the slope and particle A is 0.3. The string passes over a small smooth pulley at the top of the slope and particle B hangs vertically below the pulley. The length of the slope is 4 m and the length of the string is 3 m. Particle B is 1 m above the ground. Particle A is released and moves up the slope. When particle B reaches the ground the string is cut. Show that particle A does not reach the pulley. [11]

10 Particles A and B, of masses 0.2 kg and 0.45 kg respectively, are connected by a light inextensible string of length 2.8 m. The string passes over a small smooth pulley at the edge of a rough horizontal surface, which is 2 m above the floor. Particle A is held in contact with the surface at a distance of 2.1 m from the pulley and particle B hangs freely (see diagram). The coefficient of friction between A and the surface is 0.3. Particle A is released and the system begins to move.

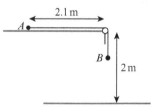

  i   Find the acceleration of the particles and show that the speed of B immediately before it hits the floor is 3.95 m s^{-1}, correct to 3 significant figures. [7]

  ii Given that B remains on the floor, find the speed with which A reaches the pulley. [4]

Cambridge International AS & A Level Mathematics 9709 Paper 42 Q6 June 2010

11

Particles A and B, of masses $0.3\,\text{kg}$ and $0.7\,\text{kg}$ respectively, are attached to the ends of a light inextensible string. Particle A is held at rest on a rough horizontal table with the string passing over a smooth pulley fixed at the edge of the table. The coefficient of friction between A and the table is 0.2. Particle B hangs vertically below the pulley at a height of $0.5\,\text{m}$ above the floor (see diagram).
The system is released from rest and $0.25\,\text{s}$ later the string breaks. A does not reach the pulley in the subsequent motion. Find

    i   the speed of B immediately before it hits the floor, [9]

    ii the total distance travelled by A. [3]

Cambridge International AS & A Level Mathematics 9709 Paper 41 Q7 June 2015

12 A slope is inclined at $30°$ to the horizontal. A small box A, of mass $2\,\text{kg}$, is held on the slope. Box A is attached to one end of a light inextensible string. The string passes over a small smooth pulley, P_1, fixed at the top of the slope and then passes under a small smooth pulley, P_2, fixed near ground level. The other end of the string is attached to a small box B, of mass $2\,\text{kg}$, at rest on the ground. The ground is horizontal and the portion of the string between pulley P_2 and box B is horizontal (see diagram). The coefficient of friction between box A and the slope is 0.2 and the coefficient of friction between box B and the ground is μ.

The distance from the bottom of the slope to A is $1\,\text{m}$ and the distance from pulley P_2 to B is $0.6\,\text{m}$. Box A is released and the system begins to move. It takes $1\,\text{s}$ for box B to reach pulley P_2.

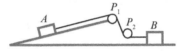

    a Find the speed of box B just before it hits the pulley. [2]

    b Show that the tension in the string is $4.14\,\text{N}$, to 3 significant figures. [3]

    c Find the value of μ. [3]

When box B hits the pulley, the string breaks.

    d Find the speed of box A just before it reaches the bottom of the slope. [6]

Chapter 6
General motion in a straight line

In this chapter you will learn how to:

- use differentiation to calculate velocity when displacement is given as a function of time
- use differentiation to calculate acceleration when velocity is given as a function of time
- use integration to find displacement when velocity is given as a function of time
- use integration to find velocity when acceleration is given as a function of time.

Chapter 6: General motion in a straight line

PREREQUISITE KNOWLEDGE

Where it comes from	What you should be able to do	Check your skills

Chapter 1	Calculate velocity and displacement when acceleration is constant.	1 A particle travels in a straight line. It has initial velocity of $8\,\text{m s}^{-1}$ and a constant acceleration of $2\,\text{m s}^{-2}$.
		  a Find the speed of the particle after it has been travelling for $3\,\text{s}$.
		  b Calculate the distance travelled in the first $3\,\text{s}$.
		  c Calculate the time it takes for the particle to travel $65\,\text{m}$.

Pure Mathematics 1	Differentiate expressions of the form kx^n, where n may be a fraction or may be negative.
	
	Integrate expressions of the form $kx^n\,(n \neq -1)$.
	
	Evaluate definite and indefinite integrals.
		2 $y = 5x^3 - 60x + 2$
		  a Find $\dfrac{dy}{dx}$ and find the coordinates of the stationary points of y.
		  b Find $\int y\,dx$ and $\int_0^2 y\,dx$.

How do objects move when acceleration is not constant?

When someone drives a car through a town centre, they will need to brake and accelerate to deal with traffic conditions. You might want to know the speed of the car at any moment or how quickly the car is speeding up or slowing down.

In Chapter 1 you learnt about displacement, velocity and acceleration and how these are connected when acceleration is constant, however, for the driver in town, acceleration is not likely to be constant. For example, the driver may gradually increase the braking force, so the rate of deceleration gradually increases, as the car comes to a stop at traffic lights.

In this chapter you will learn that if acceleration can be written as a function of time, you can use calculus (differentiation and integration) to deal with variable acceleration.

Other examples of non-constant acceleration include objects attached to springs, objects moving in circles, pendulums, and rockets leaving the Earth's surface (where air resistance, driving force and gravity are all changing during the motion). In many of these situations the direction of motion is changing, however, in this chapter you will just consider one-dimensional motion.

6.1 Velocity as the derivative of displacement with respect to time

On a displacement–time graph the velocity is represented by the gradient of the graph. This is true whether the graph is made of straight lines or curves.

EXPLORE 6.1

A car travels in a straight line. The diagram shows the displacement–time graph for the car as it slows down to approach a red traffic light.

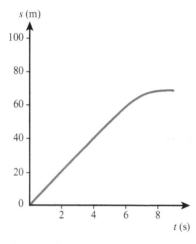

Use the graph to estimate the velocity of the car:

a between $t = 1\,\text{s}$ and $t = 3\,\text{s}$
b at $t = 2\,\text{s}$
c between $t = 4\,\text{s}$ and $t = 9\,\text{s}$
d at $t = 6\,\text{s}$.

REWIND

Look back to Chapter 1, Section 1.4 if you need a reminder of displacement–time graphs.

Average velocity is found by $\dfrac{\text{change in displacement}}{\text{change in time}}$. If these changes are small; for example, if the change in displacement is δs during a time δt, then the average velocity over this small time is $\dfrac{\delta s}{\delta t}$.

As δt gets smaller, $\dfrac{\delta s}{\delta t}$ approaches the limit $\dfrac{ds}{dt}$, which is the derivative of displacement s with respect to time t. We call this the **instantaneous velocity** or simply the velocity of the object. The velocity is represented by the gradient of a displacement–time graph, whether or not it is a straight line. If you know the displacement as a function of time, you can differentiate with respect to time to find the velocity at any instant.

REWIND

You saw in Pure Mathematics 1, Chapter 7, that the derivative of y with respect to x gives the gradient of the graph of y against x at a point.

WEB LINK

You may want to have a go at the resource *Walk-sorting* at the *Introducing Calculus* station on the Underground Mathematics website.

KEY POINT 6.1

Velocity is the rate of change of displacement and is the derivative of displacement with respect to time:

$$v = \dfrac{ds}{dt}$$

Chapter 6: General motion in a straight line

WORKED EXAMPLE 6.1

A particle moves in a straight line so that its displacement, s m, at time t s is given by $s = t^3 - 14t$. Find an expression for its velocity at time t.

Answer

$s = t^3 - 14t$ — Differentiate this with respect to time to get velocity.

$v = \dfrac{ds}{dt}$

$= (3t^2 - 14)\,\text{m s}^{-1}$ — Remember to give units.

WORKED EXAMPLE 6.2

A ball moves in a straight line so that its displacement, s m, at time t s is given by $s = 2t^3 - 10t^2$. Find its speed when $t = 2$.

Answer

$s = 2t^3 - 10t^2$ — Differentiate this with respect to time to get velocity.

$v = \dfrac{ds}{dt}$

$= 6t^2 - 20t$

When $t = 2$, $v = 24 - 40 = -16$ — Substitute $t = 2$.

So speed $= 16\,\text{m s}^{-1}$. — Speed $= |\text{velocity}|$

Remember to give units.

WORKED EXAMPLE 6.3

A particle moves forwards and backwards along a straight line so that its displacement, s metres from the initial position, at time t seconds is given by $s = 2t^3 - 12t^2 + 18t$. Find the distance that it travels in the first 5 s.

Answer

$s = 2t^3 - 12t^2 + 18t$ — You can calculate the displacement when $t = 5$, but this is not necessarily the same as the distance travelled.

Displacement when $t = 5$ is $250 - 300 + 90 = 40$ m.

Although the particle moves in a straight line, it is moving backwards as well as forwards along the line.

Displacement measures how far an object is from the start; distance measures how far it has travelled to get to that position.

$v = \dfrac{ds}{dt}$

$= 6t^2 - 24t + 18$

Differentiate s with respect to t to get an expression for v.

Stationary when $v = 0$

$6t^2 - 24t + 18 = 0$

$6(t-1)(t-3) = 0$

$t = 1$ or $t = 3$

Set $v = 0$ to find any times when the particle is stationary.

These are the times when the particle momentarily stops, before possibly changing direction.

Displacement when $t = 1$ is
$2 - 12 + 18 = 8$ m.

Set $t = 1$ and $t = 3$ in $s = 2t^3 - 12t^2 + 18t$ to find the displacement at each stationary point.

Displacement when $t = 3$ is
$54 - 108 + 54 = 0$ m.

Distance travelled $= 8 + 8 + 40$
$= 56$ m

Particle travels from $s = 0$ to $s = 8$, then from $s = 8$ back to $s = 0$, and finally from $s = 0$ to $s = 40$.

$t = 0 \quad t = 1$

$t = 3$

> **TIP**
>
> Sometimes an object travels back the way it came. It is useful to consider when it is momentarily at rest (the times when $v = 0$) as these are the times when it could change direction.

DID YOU KNOW?

In 1684 Edmund Halley asked Isaac Newton what orbit would be followed by a body under an inverse square force. Newton replied immediately that it would be an ellipse and that he could prove this using his new methods (essentially calculus methods applied to situations from mechanics).

Halley then encouraged Newton to write up his work and in 1687 Newton published his *Philosophiae Naturalis Principia Mathematica* (which is usually called the *Principia*). In the *Principia* Newton analysed the motion of bodies, including orbits, projectiles, pendulums and objects in free-fall near the surface of the Earth.

EXERCISE 6A

 1 A particle moves along the x-axis so that its x-coordinate at time t s is given by s m, where $s = 20t - 4$. Show that the particle is moving with constant velocity and find the value of this velocity.

2 A particle moves in a straight line so that its displacement, s m from the start position, at time t s is given by $s = 2t^4 + 3t^2 + 10t$. Find its velocity when $t = 2$.

 3 A tennis ball travels vertically upwards in a straight line. The displacement of the ball, measured from the initial position in metres, is modelled as $s = -5t^2 + 20t$, where t is the time from the start, in seconds.

    a What modelling assumptions have been made?

> **TIP**
>
> Throughout this chapter you may be able to check numerical derivatives on your calculator. If you have an equation solver on your calculator, you may also find this helpful. However, you will need to show full working in the examination.

Find the speed of the ball:

b when $t = 0$

c when $t = 2$.

4 A child on a fairground ride moves in a straight line. The position of the child, measured from the start, at time t s is given by $s = 0.5t^4 - t^2$ for $0 \leq t \leq 2$.

Find the speed of the child:

a when $t = 0$ **b** when $t = 1$ **c** when $t = 2$.

5 The position of a particle as it moves along a line is modelled as $s = 3 + 4t - t^2$, where s is the displacement, in metres, from a fixed point O and t is the time, in seconds, from the start.

a Show that the particle started 3 m from O.

b Find how far the particle is from O when it is instantaneously at rest.

6 A tennis ball is projected vertically upwards. The vertical displacement of the ball, in metres, from the point of projection at time t seconds is given by $s = -5t^2 + 8t$.

a Find the time when the ball returns to its starting point.

b Find the displacement when the ball is momentarily stationary.

7 A small stone is dropped into a lake. The stone descends vertically so that t s after entering the water it is s m below the surface of the water, where $s = 4t^2 - \sqrt{2t^3}$. The stone lands at the bottom of the lake with speed 13 m s^{-1}.

a Show that the stone takes 2 s to reach the bottom of the lake.

b Work out the depth of the lake at the point where the stone lands.

PS 8 In a drag race, two cars, A and B, line up side by side at the start point. When the starting flag is waved, both cars are driven as fast as possible in a straight line. The first car to cross the finish line is the winning car. At time t s from when the cars start to move, the distance travelled by car A is given by $s = 4t + t^2$. Car A takes 16 s to reach the finish line.

a Work out the distance from the start to the finish.

b Find the speed of car A when it crosses the finish line.

At time t s from when the cars start to move, the distance travelled by car B is given by $s = 1.2t^2$.

c Find the speed of car B when it crosses the finish line.

d When the winning car crosses the finish line, how far behind it is the other car?

PS 9 A burglar moves along a straight corridor from one door to the next door. At time t s his distance, s m, from the door of the first room is given by $s = 1.8t^2 - 0.3t^4$. He starts and finishes with speed $v = 0$ m s^{-1}. Find the distance between the two doors.

10 At time ts after jumping from a plane, the distance fallen by a parachutist is modelled as s m, where

$$s = \begin{cases} 5t^2 & : 0 \leq t \leq 4 \\ A\sqrt{t} + Bt & : 4 < t \leq 25 \\ Ct + 30 & : 25 < t \leq 50 \end{cases}$$

A, B and C are constants.

a Explain why $A + 2B = 40$.

The parachute is opened at $t = 4$ and the speed of the parachutist is immediately reduced by x m s^{-1}.

b Show that $0.25A + B = 40 - x$.

At $t = 25$ the speed of the parachute becomes constant.

c Write down two equations that connect A, B and C.

d Find the value of x.

11 A particle moves forwards and backwards along a straight line. The displacement of the particle, s m, from its initial position O is given by $s = -0.6t^4 + 2.4t^3 - 3.6t^2 + 2.4t$ for $0 < t < 4$, where t s is the time for which the particle has been travelling.

a Show that the particle starts moving along the line in the positive direction.

The particle comes to instantaneous rest at point A, returns to pass through O and continues to point B, where the distance OB is the same as the distance OA.

b Find the speed of the particle when it is at B.

(Note: you will need an equation solver for this question. You will not be allowed an equation solver in the examination.)

12 A robot moves along a straight line for 10 s. The displacement of the robot, s m, from its initial position is given by $s = -0.01t^4 + 0.2t^3 - 1.32t^2 + 3.2t$, where t is measured in seconds and $0 < t < 10$.

a Show that the robot is stationary when $t = 2$, $t = 5$ and $t = 8$.

b Find the distance that the robot travels in the 10 s.

6.2 Acceleration as the derivative of velocity with respect to time

You learnt in Chapter 1 that average acceleration = $\dfrac{\text{change in velocity}}{\text{change in time}}$. If these changes are small, for example, the change in velocity is δv during a time δt, then the average acceleration over this small time is $\dfrac{\delta v}{\delta t}$. As δt gets smaller, $\dfrac{\delta v}{\delta t}$ approaches the limit $\dfrac{dv}{dt}$, which is the derivative of velocity with respect to time. We call this the **instantaneous acceleration** or just the acceleration of the object. This means that the acceleration is represented by the gradient of a velocity–time graph, whether or not the graph is made up of straight lines. If you know the velocity as a function of time, you can differentiate with respect to time to find the acceleration at any instant.

 REWIND

Look back to Chapter 1, Section 1.5 if you need a reminder of velocity–time graphs and acceleration.

Chapter 6: General motion in a straight line

Acceleration is a vector quantity, like velocity. Although the magnitude of the velocity is called speed, there is no term for the magnitude of the acceleration. In this chapter you will consider only examples of one-dimensional motion along a line. One direction along the line will be the positive direction and the other will be the negative direction.

> **KEY POINT 6.2**
>
> Acceleration is the rate of change of velocity and is the derivative of velocity with respect to time. Acceleration is the second derivative of displacement with respect to time.
>
> $$a = \frac{dv}{dt} = \frac{d^2s}{dt^2}$$

EXPLORE 6.2

Positive acceleration means that the velocity is increasing. In each of the following velocity–time graphs the acceleration is positive. Give a possible description of the motion in each case. What other shapes could a velocity–time graph have for there to be positive acceleration?

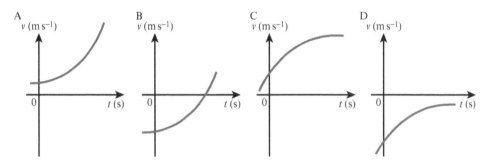

Negative acceleration (deceleration) means that the velocity is decreasing. In each of the following velocity–time graphs the acceleration is negative. Give a possible description of the motion in each case. What other shapes could a velocity–time graph have for there to be negative acceleration?

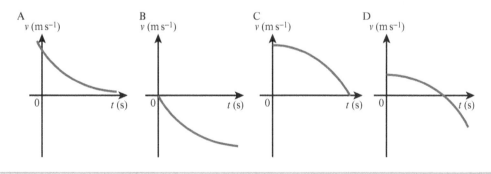

> **TIP**
>
> The sign of the velocity determines the direction of travel.
>
> If the velocity is positive then when it increases the object speeds up, but if the velocity is negative then when it increases it becomes less negative and the object slows down but in the negative direction.
>
> If the velocity is positive then when it decreases the object slows down, but if the velocity is negative then when it decreases it becomes more negative and the object speeds up but in the negative direction.

Cambridge International AS & A Level Mathematics: Mechanics

WORKED EXAMPLE 6.4

A particle moves in a straight line so that its velocity, $v\,\text{m s}^{-1}$, at time t s is given by

$v = t^2 - 4t$.

Find an expression for its acceleration at time t.

Answer

$v = t^2 - 4t$ — Differentiate this with respect to time to get acceleration.

$a = \dfrac{dv}{dt}$

$= (2t - 4)\,\text{m s}^{-2}$ — Remember to give units.

WORKED EXAMPLE 6.5

A car moves in a straight line so that its velocity, $v\,\text{m s}^{-1}$, at time t s is given by $v = 5t^2 - t^3$, for $0 < t < 4$. Find its acceleration when $t = 2$.

Answer

$v = 5t^2 - t^3$ — Differentiate this with respect to time to get acceleration.

$a = \dfrac{dv}{dt}$

$= 10t - 3t^2$

When $t = 2$: — Substitute $t = 2$.

$a = 20 - 12$

$= 8$

So acceleration $= 8\,\text{m s}^{-2}$. — Remember to give units.

WORKED EXAMPLE 6.6

A particle moves in a straight line so that its displacement, s m, at time t s $(0 \leq t \leq 10)$ is given by

$s = \dfrac{1}{3}t^3 - 6t^2 + 15t$.

  a Sketch the shape of the velocity–time graph for the particle.

  b Hence, find the maximum speed of the particle.

Answer

a $s = \dfrac{1}{3}t^3 - 6t^2 + 15t$ .... First differentiate to find v as a function of t.

$v = \dfrac{ds}{dt}$

$= t^2 - 12t + 15$

v (m s⁻¹) graph .... The velocity–time graph is a parabola.

You could use a graphic calculator or a graph-drawing package to check the shape of the graph.

[Graph showing parabola starting at v=15 when t=0, crossing t-axis, reaching minimum around t=6, and rising back toward t=10]

b Speed is the magnitude of v, so to find the maximum speed we need to find the maximum positive value and the minimum negative value of v.

.... From the graph you can see that the minimum value occurs at the turning point and the maximum value occurs when $t = 0$.

$v = t^2 - 12t + 15$ .... Differentiate again to find the acceleration.

$a = \dfrac{dv}{dt}$

$= 2t - 12$

$\dfrac{dv}{dt} = 0$ when $2t - 12 = 0$ so $t = 6$. .... This is the time when the velocity–time graph has its turning point.

Alternatively, complete the square for the quadratic to get

$v = (t - 6)^2 - 21$

When $t = 6$: .... This is v at the minimum turning point on the graph. The minimum value of v is –21 (and the speed at this point is $21\,\text{m s}^{-1}$).

$v = 36 - 72 + 15$

$= -21$

When $t = 0$: .... These are the values of v at the end points of the graph. (We only really need to work out the value when $t = 0$ because we can see from the graph this is where the velocity is greatest.)

$v = 0 - 0 + 15$

$= 15$

When $t = 10$:

$v = 100 - 120 + 15$

$= -5$

The maximum value of v is 15 (and the speed at this time is $15\,\text{m s}^{-1}$).

Hence, the maximum speed $= 21\,\text{m s}^{-1}$. .... Speed $= |\text{velocity}|$

EXERCISE 6B

1. A particle moves in a straight line so that its velocity, v m s^{-1}, at time t s ($t > 0$) is given by $v = 5t^2 - 20$. Find the time when the particle is stationary.

2. A particle moves in a straight line so that its velocity, v m s^{-1}, at time t s is given by $v = t^3 + 3t^2 - 8t + 1$. Find its acceleration when $t = 2$.

M 3. A tennis ball travels vertically upwards in a straight line. The velocity of the ball in the upwards direction, v m s^{-1}, at time t s is given by $v = 20 - 10t$.

   a. Find the acceleration of the ball.

   b. Interpret your result.

4. A car moves in a straight line. The velocity of the car, v m s^{-1}, at time t s is given by $v = 5t + 0.5t^2$ for $0 \leq t \leq 2$. Find the acceleration of the car:

   a. when $t = 0$

   b. when $t = 1$

   c. when $t = 2$.

P 5. A boy runs in a straight line. The velocity of the boy, v m s^{-1}, at time t s is given by:

$$v = \begin{cases} 2 + 3t &: 0 \leq t \leq 1 \\ 6 - t^2 &: 1 \leq t \leq 1.2 \\ 5 - \dfrac{k}{30}t &: 1.2 \leq t \leq T \end{cases} \text{ for } 0 \leq t \leq T$$

   The boy has velocity 0 m s^{-1} at time T.

   a. Work out the value of k.

   b. Show that T is just under 14 s.

   c. Work out the average acceleration of the boy over the period 0–T.

   d. Show that there is no time at which the acceleration of the boy is the same as his average acceleration.

6. The motion of a cat, moving along a straight line, is modelled as $s = 6t - t^2$ for small values of t, where s is measured in metres and t in seconds.

   a. Find an expression for the velocity of the cat as a function of time.

   b. Describe the motion of the cat.

   c. When does the cat come to momentary rest?

   d. Find the acceleration of the cat.

7. A particle moves forwards and backwards along a straight line. The velocity of the particle, v m s^{-1}, at time t s is given by $v = -t^3 + 75t$ for $0 < t < 10$. Find the maximum speed of the particle.

8 A robot moves along a straight line for 3s. The displacement of the robot, s m, from its initial position is given by $s = At^3 + Bt^2 + Ct$ for constants A, B and C, where t is measured in seconds and $0 < t < 3$. The robot starts with velocity $2\,\text{m s}^{-1}$. It travels 6 m before coming to rest at time $t = 3$. Work out the values of A, B and C.

9 A particle moves along a straight line. The displacement of the particle, s m, at time t s is given by $s = 4t^2 - 0.25t^3$ for $0 < t < 12$.

   a Find the maximum speed of the particle.

   b Work out the difference between the time when the speed is greatest and the time when the speed is least.

10 A ball moves in a straight line. The velocity of the ball, $v\,\text{m s}^{-1}$, at time t s is given by $v = 5 + 4t - t^2$ for $0 < t < 5$.

   a Write down the initial velocity of the ball.

   b Work out the maximum velocity of the ball.

   c Find the average acceleration between the start and the end of the motion.

11 A train is travelling in a straight line. Alice is sitting on the train and is using her mobile phone, which is being tracked. The position of the phone, measured from when the tracking began, is given by $s = t^2 - 2t^3 + 75t$, where s is measured in km and t is measured in hours. The phone is tracked for 2 hours, so $0 < t < 2$. Initially the train speeds up but then it slows down again.

   a After how long is the train travelling at its fastest speed?

   b Find the maximum velocity.

   c Find how far the train travels before it starts to slow down.

6.3 Displacement as the integral of velocity with respect to time

You can differentiate displacement (as a function of time) to find velocity. Reversing this means that if you integrate velocity (with respect to time) you will get a function for the displacement.

Displacement measures how far the object is from an origin. In this chapter the origin will be at the initial position for the motion, unless a question states otherwise. This means that the displacement, s, will usually be 0 when $t = 0$ (and the constant of integration will usually be 0, unless the velocity function is made up of different pieces).

> **TIP**
>
> You know that integrating a function gives the area under its graph and that the area under a velocity–time graph gives the displacement. So, integrating velocity with respect to time will give displacement.

KEY POINT 6.3

$$v = \frac{ds}{dt} \quad \text{so} \quad s = \int v \, dt$$

This is the velocity–time graph for a particle:

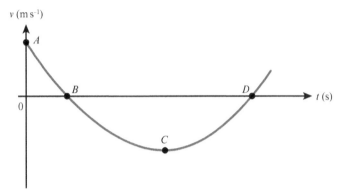

You can interpret the motion as follows:

- The particle starts at A with positive velocity.
- Velocity is initially decreasing so the particle is slowing down. However, the velocity is positive so displacement is increasing and the particle is moving away from the starting point.
- When the graph crosses the horizontal axis (at B) the particle has velocity $0\,\text{m s}^{-1}$ and is momentarily at rest.
- The particle's velocity continues to decrease and is now negative, so it is now travelling in the opposite direction (back towards where it started from).
- The velocity decreases to a minimum (at C) and then starts to increase, but remains negative. This means that the particle continues to travel in the negative direction (towards the start point) but is slowing down.
- When the graph again crosses the horizontal axis (at D), the particle is once more momentarily at rest before it changes direction. The velocity continues to increase, but now in a positive direction.

Suppose that the equation for the velocity is:

$$v = t^2 - 13t + 22 = (t-2)(t-11)$$

then the graph crosses the horizontal axis ($v = 0$) when $t = 2$ and when $t = 11$.

The particle starts with velocity $22\,\text{m s}^{-1}$. For $0 \leqslant t \leqslant 2$ the particle moves in the positive direction, for $2 \leqslant t \leqslant 11$ it moves in the negative direction, and for $t \geqslant 11$ it moves in the positive direction again.

To find the displacement from the starting point, you integrate the equation for v.

$$s = \int v \, dt = \int (t^2 - 13t + 22) \, dt = \frac{1}{3}t^3 - \frac{13}{2}t^2 + 22t$$

(the constant of integration will be 0 because $s = 0$ when $t = 0$).

The following table shows the displacement for some values of t.

t (s)	0	1	2	3	4	5	6	7	8	9	10	11	12
s (m)	0	15.8	20.7	16.5	5.3	−10.8	−30.0	−50.2	−69.3	−85.5	−96.7	−101	−96.0

You can see that the displacement increases as the particle travels away from the start for the first 2 seconds. It then decreases as the particle returns the way it came and passes through the start point (when displacement is zero). The displacement continues to decrease until the particle is a distance of 101 from the start, but in the negative direction. The direction then changes again as the particle travels back towards the start.

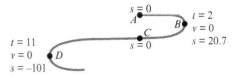

Displacement and distance travelled are not necessarily the same thing. The displacement at time t is the position of the object from the origin at that time, but the distance travelled at time t is the sum of the distances travelled in the positive and negative directions up to that time. For example, when $t = 3$ the displacement of the particle is 16.5. However, the object has travelled 20.7 in the positive direction and $20.7 - 16.5 = 4.2$ in the negative direction, so the total distance travelled is $20.7 + 4.2 = 24.9$.

The distance travelled is the area between the curve and the horizontal axis, but in this example some of the curve is below the horizontal axis. To find the total distance travelled we need to find the times when the graph crosses the axis and then integrate the parts above and below the axis separately to find the distances travelled in the positive and negative directions. The total distance travelled is the sum of these distances.

> **TIP**
>
> Unless you are told otherwise, in work involving calculus the displacement will be measured from the initial position, so s will be 0 when $t = 0$.

WORKED EXAMPLE 6.7

A particle moves in a straight line so that its velocity, $v\,\text{m s}^{-1}$, at time t s is given by $v = t^3 - 4t$. Find an expression for its displacement from the initial position at time t.

Answer

$v = t^3 - 4t$ .... Integrate this with respect to time to get displacement.

$s = \int (t^3 - 4t)\,dt$

$= \dfrac{1}{4}t^4 - 2t^2 + c$

But $s = 0$ when $t = 0$, so $c = 0$. .... Displacement is measured from the initial position, so $s = 0$ when $t = 0$.

$s = \dfrac{1}{4}t^4 - 2t^2$ m

Alternatively:

$s = \displaystyle\int_0^t (t^3 - 4t)\,dt$ .... Here we are using t as a variable within the integration and also as a constant in the limit.

$= \left[\dfrac{1}{4}t^4 - 2t^2\right]_0^t$

$= \dfrac{1}{4}t^4 - 2t^2$ m

WORKED EXAMPLE 6.8

A particle moves in a straight line so that its velocity, $v\,\text{m s}^{-1}$, at time t s is given by $v = t^3 - 5t^2$.

a Find the displacement of the particle after 5 s.

b Sketch the velocity–time graph for $0 \leqslant t \leqslant 10$.

c Work out the distance that the particle travels in the first 10 s.

d Find the time when the particle passes through the start position.

Answer

a $v = t^3 - 5t^2$ .... Integrate this with respect to time to get displacement.

In the first 5 seconds the displacement is .... The limits are $t = 0$ and $t = 5$.

$$s = \int_0^5 (t^3 - 5t^2)\,dt$$

$$= \left[\frac{1}{4}t^4 - \frac{5}{3}t^3\right]_0^5$$

$$= -52.1\,\text{m}$$ .... Displacement can be positive or negative.

b The particle is at instantaneous rest when $t^3 - 5t^2 = 0$.

$t^2(t - 5) = 0$ .... Bring out common factor t^2.
$t = 0$ corresponds to the start of the motion.

$t = 0$ or $t = 5$

Hence, it is at instantaneous rest when $t = 5$.

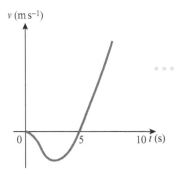

$t = 0$ is a repeated root so the graph has a stationary point at $t = 0$.
v is negative for small values of t.
The graph is part of a cubic curve.

c The distance travelled is the distance travelled from $t = 0$ to $t = 5$, added to the distance travelled from $t = 5$ to $t = 10$. .... The particle is momentarily at rest when $t = 5$.

$$\int_0^5 (t^3 - 5t^2) \, dt = \left[\frac{1}{4} t^4 - \frac{5}{3} t^3 \right]_0^5$$

$$= -52.08$$

.... From $t = 0$ to $t = 5$ the particle moves 52.08 m (in the negative direction).

$$\int_5^{10} (t^3 - 5t^2) \, dt = \left[\frac{1}{4} t^4 - \frac{5}{3} t^3 \right]_5^{10}$$

$$= 885.42$$

.... From $t = 5$ to $t = 10$ the particle moves 885.42 m (in the positive direction).

Total distance travelled $= 52.08 + 885.42$
$= 938 \text{ m}$

Alternative method:

$$s = \int (t^3 - 5t^2) \, dt$$

$$= \frac{1}{4} t^4 - \frac{5}{3} t^3 + c$$

.... $s = 0$ when $t = 0$, so $c = 0$.

$$= \frac{1}{4} t^4 - \frac{5}{3} t^3$$

When $t = 5$, $s = -52.08$.

When $t = 10$, $s = 833.33$.

.... Particle moves 52.08 m in the negative direction, then turns and travels through the start position to finish 833.33 m from the start in the positive direction.

Total distance travelled $= 52.08 + 52.08 + 833.33$ .... Particle travels 52.08 twice, but this rounds up to 104.17 when exact values are used.
$= 938 \text{ m}$

d Displacement $= 0$ when $\frac{1}{4} t^4 - \frac{5}{3} t^3 = 0$. .... Set $s = 0$.

Bring out common factor t^3.

$$\frac{1}{12} t^3 (3t - 20) = 0$$

$t = 0$ or $\frac{20}{3}$

Particle passes through start position after $6\frac{2}{3}$ seconds.

WORKED EXAMPLE 6.9

A circus performer rides a unicycle along a stretched tightrope. The tightrope is modelled as a straight horizontal line and the velocity, $v \, \text{m s}^{-1}$, is modelled as:

$v = 1$ for $0 \leq t \leq 2$

$v = 2t - 3$ for $2 \leq t \leq 4$

$v = 7 - \sqrt{t}$ for $4 \leq t \leq T$

where t is the time, in seconds, from when the performer starts to cycle along the tightrope.

The performer stops at time T s, having just reached the other end of the tightrope.

a Find the value of T.

b Calculate how far the performer cycles.

Answer

a $7 - \sqrt{T} = 0$ .... $v = 0$ at time T.

   $T = 49$ .... Performer stops after 49 s.

b For $0 \leq t \leq 2$: $s = t + c_1$ .... Integrate v with respect to t.

   $s = 0$ when $t = 0$, so $c_1 = 0$. .... s is measured from the start of the tightrope.

   Hence, $s = t$.

   So when $t = 2$ s, the distance cycled is $s = 2$ m.

   For $2 \leq t \leq 4$: $s = t^2 - 3t + c_2$ .... Integrate $2t - 3$ with respect to t.

   $s = 2$ when $t = 2$, so $c_2 = 4$.

   Hence, $s = t^2 - 3t + 4$.

   So when $t = 4$ s, the total distance cycled is $s = 8$ m.

   For $4 \leq t \leq 49$: $s = 7t - \frac{2}{3}t^{1.5} + c_3$ .... Integrate $7 - t^{0.5}$ with respect to t.

   $s = 8$ when $t = 4$, so $c_3 = -14\frac{2}{3}$.

   Hence, $s = 7t - \frac{2}{3}t^{1.5} - 14\frac{2}{3}$.

   So when $t = 49$ s the total distance cycled is $s = 99\frac{2}{3}$ m.

Alternatively, we can integrate each of the expressions for v with limits, and add the distances obtained.

$$\int_0^2 1 \, dt = [t]_0^2$$
$$= 2$$

$$\int_2^4 (2t - 3) \, dt = [t^2 - 3t]_2^4$$
$$= 6$$

$$\int_4^{49} (7 - t^{0.5}) \, dt = \left[7t - \frac{2}{3}t^{1.5}\right]_4^{49}$$
$$= 91\tfrac{2}{3}$$

The total distance travelled $= 2 + 6 + 91\tfrac{2}{3}$
$$= 99\tfrac{2}{3} \text{ m}$$

WORKED EXAMPLE 6.10

A particle moves in a straight line so that its velocity, $v \, \text{m s}^{-1}$, at time t seconds after it starts to move is given by $v = u + at$, where u and a are constants. Find the displacement after t s.

Answer

$v = u + at$ — Integrate this with respect to time to get displacement.

$s = \int (u + at) \, dt$ — u and a are constants.

$= ut + \frac{1}{2}at^2 + c$

But $s = 0$ when $t = 0$, so $c = 0$.

Hence, $s = ut + \frac{1}{2}at^2$. — You should recognise this as one of the constant acceleration formulae from Chapter 1.

EXERCISE 6C

1. A particle moves in a straight line so that its velocity, $v\,\text{m s}^{-1}$, at time $t\,\text{s}\,(t>0)$ is given by $v = 5t^2 - 20$. Find the displacement of the particle when it is stationary.

2. A particle moves in a straight line so that its velocity, $v\,\text{m s}^{-1}$, at time $t\,\text{s}$ is given by $v = t^3 + 3t^2 - 8t + 1$. Find its displacement when $t = 3$.

3. A tennis ball is hit vertically upwards. The upward velocity, $v\,\text{m s}^{-1}$, of the ball at time $t\,\text{s}$ is given by $v = 20 - 10t$. Find the upward displacement of the ball, from the initial position, when $v = 0$.

4. A speed skater moves in a straight line with velocity $v\,\text{m s}^{-1}$ at time $t\,\text{s}$, given by $v = 2t + t^2$, for $0 \leqslant t \leqslant 2$. Find the displacement of the skater:

   a when $t = 0$

   b when $t = 1$

   c when $t = 2$.

5. A small stone is dropped into a well. It falls down the well, from rest, with no resistance for 2 s. It then hits the surface of the water and continues to fall vertically through the water until it reaches the bottom of the well. In the water the downwards velocity of the stone, $v\,\text{m s}^{-1}$, is given by $v = 20 - t$, where t is the time, in seconds, measured from when the stone hits the surface of the water. The stone takes 2.5 s in total to reach the bottom of the well.

   a Calculate the depth of the well.

   b If air resistance is taken into account, would you expect the depth of the well to be greater or smaller than your answer from part **a**?

6. A ball bearing is fired vertically upwards in a straight line through a tub of butter. The upward velocity of the ball bearing is given by $v = 13 - 10t - 3t^2$ cm s^{-1}, where t is the time from when it was fired upwards.

   a Find the time when the ball bearing comes momentarily to rest.

   b Find how far the ball bearing has travelled upwards at this time.

   The ball bearing then falls downwards through the hole it has made in the butter. The downward velocity of the ball bearing is given by $v = 10T$ cm s^{-1}, where T is the time from when it was momentarily at rest.

   c Find the time that the ball bearing takes (from when it was momentarily at rest) to fall to its original position.

   d What assumptions have been made in the model used in part **c**, and how could the model be improved?

7. A particle moves on a straight line. The velocity of the particle, $v\,\text{m s}^{-1}$, at time $t\,\text{s}$ is given by $v = -t^3 + 9t\,\text{m s}^{-1}$ for $0 < t < 5$.

   a Find the displacement of the particle, from its original position, when $t = 5$.

   b Work out the distance that the particle travels from $t = 0$ to $t = 5$.

8. A car moves in a straight line with velocity $v\,\text{m s}^{-1}$ at time $t\,\text{s}$, given by $v = 16 - 0.5t^{1.5} + t$ for $0 \leqslant t \leqslant 25$.

   a Find the displacement of the car, from its original position, when $t = 25$.

   b Work out the distance that the car travels from $t = 0$ to $t = 25$. (Note: you will need an equation solver for this question. In the examination you will only be asked to solve the equations that can be done using algebraic methods.)

9 A ball rolls forwards and backwards in a long straight tube. The velocity, v m s^{-1}, at time t s after measurement starts is given by:

$v = 6 + t^{0.5}$ for $0 \leqslant t \leqslant 25$

$v = A - t$ for $t \geqslant 25$

for some constant A.

   a Show that $A = 36$.

   b Find the distance from the start when the ball changes direction.

10 A particle moves in a straight line, starting from rest. At time t s after the start, the velocity, v m s^{-1}, of the particle is given by:

$v = 2t$ for $0 \leqslant t \leqslant 4$

$v = 9 - (t - 5)^2$ for $4 \leqslant t \leqslant 10$

   a Find the maximum speed of the particle.

   b Work out the distance that the particle moves in the time interval $0 \leqslant t \leqslant 10$.

11 Two cars travel towards one another on the two sides of a long straight road. They start 300 m apart and each car stops when its speed is 0 m s^{-1}. The time, in seconds, after the first car starts to move is t. The velocity, v_1 m s^{-1}, of the first car is given by $v_1 = 7.5t - 0.5t^2$. The second car starts 2 s after the first, at $t = 2$. The velocity, v_2 m s^{-1}, of the second car is given by $v_2 = (t - 2)^2 - 16$.

   a Write down the initial speed of each car.

   b How far does the first car travel?

   c How far does the second car travel?

   d For what value of t are the cars alongside one another? (Note: you will need an equation solver for this question. You will not be allowed an equation solver in the examination.)

12 A car travels in a straight line, starting and finishing at rest. At time t s after the start, the velocity of the car is modelled as:

$v = t$ for $0 \leqslant t \leqslant 2$

$v = kt - 0.055(t - 2)^2$ for $2 \leqslant t \leqslant T$

   a Find the value of k.

   b Show that there is no change in the acceleration of the car at $t = 2$.

   c Find the maximum velocity of the car during its journey.

   d Find the time, T s, at which the car stops.

   e Work out the distance that the car travels from the start to the end of the journey.

6.4 Velocity as the integral of acceleration with respect to time

In Section 6.2 you saw that you can differentiate velocity as a function of time to find acceleration. Reversing this means that if you integrate acceleration with respect to time, you will get a function for the velocity.

KEY POINT 6.4

$$a = \frac{dv}{dt} \text{ so } v = \int a \, dt$$

The object is not necessarily at rest when $t = 0$, so you need to include a constant of integration. If you know the velocity at some time (which may be $t = 0$ or some other time) you can use this to find the constant of integration. An alternative strategy is to find the change in the velocity by integrating (between, say, $t = 0$ and a general time t) and adding this to the velocity at the beginning of the interval. Both of these approaches are demonstrated in Worked example 6.11.

MODELLING ASSUMPTIONS

When acceleration is not constant, is it reasonable to assume the displacement, velocity and acceleration follow a neat formula?

Forces may vary according to the speed or position of an object, or according to time. When used with Newton's second law, this creates an equation that relates one or more of these variables with acceleration. In many cases this sort of equation can be solved by integration, so the displacement, velocity and acceleration may have a neat formula. However, the integration of some functions is not possible, so sometimes you need to make approximations to allow the problem to be solved.

In this chapter you have only looked at the formulae for displacement, velocity and acceleration, not at the situations they come from.

TIP

As in Worked example 6.6, you sometimes need to solve $a = 0$ to find a maximum or minimum velocity (a maximum or minimum turning point of the velocity–time graph). Remember that the maximum or minimum may also occur at an end point of the graph.

WORKED EXAMPLE 6.11

A particle moves in a straight line so that its acceleration, $a \text{ m s}^{-2}$, at time t s is given by $a = 8 - 4t$. It starts with velocity 2 m s^{-1}. Find an expression for its velocity at time t.

Answer

$a = 8 - 4t$ Integrate this with respect to time to get velocity.

$v = \int (8 - 4t) \, dt$

$= 8t - 2t^2 + c$

$v = 2$ when $t = 0$, so $c = 2$.   Initial velocity $= 2 \text{ m s}^{-1}$.

So $v = 8t - 2t^2 + 2$.

Alternatively:

$$v = 2 + \int_0^t (8 - 4t)\, dt$$ .... Initial velocity + change in velocity

$$= 2 + \left[8t - 2t^2\right]_0^t$$

So $v = 8t - 2t^2 + 2$.

WORKED EXAMPLE 6.12

A particle moves in a straight line so that its acceleration, a m s^{-2}, at time t s is given by $a = 12 - 4t - 0.6t^2$. The particle comes to rest after 5 s.

   a Find the initial velocity of the particle.

   b Find the displacement after 2 s.

Answer

   a $a = 12 - 4t - 0.6t^2$ .... Integrate this with respect to time to get velocity.

$$v = \int (12 - 4t - 0.6t^2)\, dt$$

$$= 12t - 2t^2 - 0.2t^3 + c$$

When $t = 5$, $v = 0$: .... Here we know the velocity after 5 seconds, so use $v = 0$ when $t = 5$ to find c.

$60 - 50 - 25 + c = 0$ so $c = 15$

$v = 12t - 2t^2 - 0.2t^3 + 15$

Initial velocity $= 15$ m s^{-1} .... Set $t = 0$ to find the initial velocity.

   b $s = \int v\, dt$ .... Integrate velocity with respect to time to get displacement.

$$= \int \left(12t - 2t^2 - 0.2t^3 + 15\right) dt$$

$$s = 6t^2 - \frac{2}{3}t^3 - 0.05t^4 + 15t + k$$

But $s = 0$ when $t = 0$, so $k = 0$. .... Displacement from the start is 0 at time 0.

So $s = 6t^2 - \frac{2}{3}t^3 - 0.05t^4 + 15t$.

When $t = 2$, $s = 47.9$ m.

EXPLORE 6.3

A particle moves in a straight line so that its acceleration at time t seconds is a m s^{-2}, where a is constant. The initial velocity of the particle is u m s^{-1}.

Use calculus to find the velocity of the particle as a function of t.

 WEB LINK

You may want to have a go at the resource *Thinking constantly* at the *Calculus meets functions* station on the Underground Mathematics website.

EXERCISE 6D

1. A particle moves in a straight line so that its acceleration, a m s^{-2}, at time t s is given by $a = 2t^4 + 3t^2 - 12t$. The particle starts from rest. Find its speed when $t = 2$.

2. A particle moves in a straight line so that its acceleration at time t s is given by $a = 10t - 4$ m s^{-2}. The initial velocity of the particle is 15 m s^{-1}. Find the minimum velocity of the particle in the subsequent motion.

3. A body moves in a straight line so that its acceleration, a m s^{-2}, at time t s is given by $a = 2t^3 + 6t^2 - 18t$. The body starts with velocity 5 m s^{-1}.
    a. Find the velocity when $t = 3$.
    b. Find the displacement when $t = 3$.
    c. When $t = 3$, is the body travelling towards its original position or away from it?

4. A bird moves in a straight line from point A to point B and back to point A. The bird has speed 5 m s^{-1} when it starts and moves with acceleration, given by $a = \frac{1}{12}(9t^2 - 32t - 10)$. At point B the bird has velocity 0 m s^{-1}.
    a. Show that the bird takes 2 s to travel from A to B.
    b. Find the distance from A to B.
    c. Show that the bird returns to A after about 4.25 s.
    d. Find the speed of the bird when it returns to A.

5. A block slides down a sloped surface with a varying coefficient of friction. The acceleration, a m s^{-2}, of the block (measured down the surface) is given by $a = 0.01(10t - 3)$, where t is the time in seconds. At $t = 0$ the block is at rest at the top of the surface. The block reaches the bottom of the surface with speed 2.24 m s^{-1}. How far does the block travel down the sloped surface?

6. A ball moves with acceleration given by $a = 0.01(4t + 3t^{0.5})$, where t is the time, in seconds. At $t = 1$ the ball is moving with velocity 0.5 m s^{-1}. Find the displacement of the ball between $t = 0$ and $t = 4$.

7. A robot moves in a straight line with acceleration a m s^{-2} at time t s, given by $a = 40(t - 1)^3$. The minimum velocity of the robot in the subsequent motion is 0 m s^{-1} (the velocity is never negative).
    a. Show that the robot is stationary ($v = 0$) when $t = 1$.
    b. Find the displacement of the robot, measured from the initial position, when the robot is stationary.

8 A particle starts at the origin and moves along the x-axis. The acceleration of the particle in the direction of the positive x-axis is $a = 6t - c$ for some constant c. The particle is initially stationary and it is stationary again when it is at the point with x-coordinate $= -4$. Find the value of c.

9 A goods train starts from rest at point A and moves along a straight track. The train moves with acceleration $a\,\mathrm{m\,s^{-2}}$ at time $t\,\mathrm{s}$, given by $a = 0.1t^2(6 - t)$ for $0 \leqslant t < 6$. It then moves at constant velocity for $6 \leqslant t < 156$ before decelerating uniformly to stop at point B at $t = 165$. Calculate the distance from A to B.

10 Two cars are travelling towards one another on the two sides of a long straight road. Each car stops when its speed is $0\,\mathrm{m\,s^{-1}}$. The time, in seconds, after the first car starts to move is t. The acceleration, $a_1\,\mathrm{m\,s^{-2}}$, of the first car is given by $a_1 = 5 - 2t$. The maximum velocity that the first car achieves is $26.01\,\mathrm{m\,s^{-1}}$.

   a Work out the initial velocity of the first car.

   b Find the time when the first car stops moving.

   The velocity, $v_2\,\mathrm{m\,s^{-1}}$, of the second car is given by $v_2 = t^2 - 16$.

   c How far does the second car travel?

   Initially the cars are 200 m apart.

   d Show that the cars stop before they meet.

11 An ice hockey player hits the puck so that it moves across the ice in a horizontal straight line with acceleration $a\,\mathrm{m\,s^{-2}}$ at time $t\,\mathrm{s}$, where $a = -0.03t^2$. The initial speed of the puck, along the direction of motion, is $40\,\mathrm{m\,s^{-1}}$.

   a Find the distance that the puck travels in the first 2 seconds (between $t = 0$ and $t = 2$).

   b Find the speed of the puck after 2 seconds.

   When $t = 2$ the puck is stopped by an opposing player. This player then hits the puck back the way it came, giving it an initial speed of $30\,\mathrm{m\,s^{-1}}$. The acceleration of the puck, in its direction of travel, is still given by $a = -0.03t^2$. The puck returns to its original starting point.

   c Find, to 3 significant figures, how long it takes for the puck to return to its original starting point.

   (Note: you will need an equation solver for this question. You will not be allowed an equation solver in the examination.)

12 A girl bowls a ball along a straight and horizontal skittle alley. The forces acting on the ball are its weight, the normal contact force, friction and air resistance. The coefficient of friction between the ball and the surface of the skittle alley is 0.01. The girl models the air resistance, in newtons, as $m(0.9 + 1.5t)$, where m is the mass of the ball, in kg, and t is the time, in s.

   a Show that the velocity of the ball along the skittle alley, $v\,\mathrm{m\,s^{-1}}$, is given by $v = -0.75t^2 - t + C$ for some constant C.

   The initial velocity of the ball is $8\,\mathrm{m\,s^{-1}}$. The skittle alley is $7.25\,\mathrm{m}$ long and the ball reaches the end of the skittle alley with velocity $6.25\,\mathrm{m\,s^{-1}}$.

   b Show that the ball takes 1 second to reach the end of the skittle alley.

   (Note: you will need an equation solver for this question. You will not be allowed an equation solver in the examination.)

   c Why is the model for air resistance unreasonable?

Checklist of learning and understanding

- $v = \dfrac{ds}{dt}$
- $a = \dfrac{dv}{dt} = \dfrac{d^2s}{dt^2}$
- $s = \displaystyle\int v \, dt$
- $v = \displaystyle\int a \, dt$

$$\text{displacement } s \xrightarrow{\text{differentiate with respect to } t} \text{velocity } v \xrightarrow{\text{differentiate with respect to } t} \text{acceleration } a$$

$$\text{displacement } s \xleftarrow{\text{integrate with respect to } t} \text{velocity } v \xleftarrow{\text{integrate with respect to } t} \text{acceleration } a$$

- You may be able to check numerical differentiation and numerical integrals on your calculator. If you have an equation solver on your calculator, you may also find this helpful. However, you will need to show full working in the examination.

Chapter 6: General motion in a straight line

END-OF-CHAPTER REVIEW EXERCISE 6

1. A woman on a sledge moves in a straight line across horizontal ice. Her initial velocity is $2\,\mathrm{m\,s^{-1}}$. Throughout the journey her acceleration is given by $a = -0.01t\,\mathrm{m\,s^{-2}}$, where t is the time from the start, in seconds. Find the distance that she travels before coming to rest. [4]

2. A particle moves on a straight line, starting from rest at the point O. It travels from O to A with constant acceleration $0.2\,\mathrm{m\,s^{-2}}$, taking $16\,\mathrm{s}$ to reach A. The acceleration of the particle then changes so that the velocity, in $\mathrm{m\,s^{-1}}$, is given by $v = t^{0.5} - 0.8$ for $t > 16$, where t is the time, in seconds, from the start of the motion.

   a Find the acceleration of the particle immediately after passing through A. [1]

   b Find the distance travelled from $t = 0$ to $t = 36$. [3]

3. A particle, P, starts from rest at a point O and travels in a horizontal straight line. For $0 < t < 20$, where t is the time in s, the velocity of P in $\mathrm{m\,s^{-1}}$ is given by $v = 1.2t - 0.03t^2$. When $t = 20$, P collides with another particle. After the collision the direction of travel of P is reversed. For $20 < t < 30$, the velocity of P in $\mathrm{m\,s^{-1}}$ is given by $v = 0.3t - 9$. The particle comes to rest and stops when $t = 30$.

   a Find the speed of P immediately before the collision and immediately after the collision. [2]

   b Find the total distance travelled by the particle. [2]

4. A sledge moves down a slope in a straight line. At time $t\,\mathrm{s}$ the displacement of the sledge from the start is $s\,\mathrm{m}$, where $s = 0.4t^2$ for $0 \leqslant t < 10$ and $s = \left(7t - \dfrac{100}{t} - 20\right)$ for $10 \leqslant t \leqslant 50$.

   a Find maximum velocity of the sledge. [3]

   b Show that the acceleration instantaneously reduces by $1\,\mathrm{m\,s^{-2}}$ at $t = 10$. [2]

5. A particle travels in a tube, starting from rest. The particle does not come to rest again in the subsequent motion. At time $t\,\mathrm{s}$, the particle has acceleration $a\,\mathrm{m\,s^{-2}}$, given by $a = 9000t$ until it reaches a velocity of $28\,125\,\mathrm{m\,s^{-1}}$, along the direction of the tube. How far does the particle travel while it is accelerating? [6]

6. A particle moves in a straight line, starting at time $t = 0$ and continuing until it comes to rest. While it is moving the particle has acceleration $a\,\mathrm{m\,s^{-2}}$, in the positive direction along the line. This acceleration is given by $a = 0.1 - 0.01t$, where t is the time from the start, in seconds. The particle starts with speed $4\,\mathrm{m\,s^{-1}}$ and finishes with speed $0\,\mathrm{m\,s^{-1}}$.

   a Find the maximum speed of the particle. [4]

   b Find the time when the particle comes to rest. [2]

7. A car is moving in a straight line. The acceleration, $a\,\mathrm{m\,s^{-2}}$, at time $t\,\mathrm{s}$ after the car starts to move is modelled as $a = A(1 + 4t)$ for $0 \leqslant t < 1$ and $a = B\left(30 - \dfrac{10}{t^3}\right)$ for $1 \leqslant t \leqslant 5$, where A and B are constants.

   a Show that $A = 4B$. [2]

   At $t = 5$ the velocity of the car is $31.8\,\mathrm{m\,s^{-1}}$.

   b Show that $A = 1$. [2]

   c Work out the distance travelled in the time interval $0 \leqslant t \leqslant 5$. [2]

   d By considering the acceleration–time graph at $t = 1$, criticise the model. [1]

8. A particle P moves on a straight line, starting from rest at a point O of the line. The time after P starts to move is $t\,\mathrm{s}$, and the particle moves along the line with constant acceleration $\dfrac{1}{4}\,\mathrm{m\,s^{-2}}$ until it passes through a point A at time $t = 8$. After passing through A the velocity of P is $\dfrac{1}{2}t^{\frac{2}{3}}\,\mathrm{m\,s^{-1}}$.

i Find the acceleration of P immediately after it passes through A. Hence show that the acceleration of P decreases by $\frac{1}{12}$ m s^{-2} as it passes through A. [2]

ii Find the distance moved by P from $t = 0$ to $t = 27$. [3]

Cambridge International AS & A Level Mathematics 9709 Paper 42 Q4 June 2014

9 A hockey ball is hit so that it moves in a horizontal straight line with acceleration a m s^{-2}, along the direction of travel; $a = -0.6t$, where t is the time from when the ball was hit, in seconds. The initial speed of the ball is 14 m s^{-1}.

a Find the speed of the ball when it has travelled 57.5 m. [4]

(Note: you will need an equation solver for this question. You will not be allowed an equation solver in the examination.)

b Find the distance that the ball has travelled when the ball is first momentarily stationary. [4]

c Find the value of t when the ball has travelled 40 m. [2]

(Note: you will need an equation solver for this question. You will not be allowed an equation solver in the examination.)

10 Two particles A and B start to move at the same instant from a point O. The particles move in the same direction along the same straight line. The acceleration of A at time t s after starting to move is a m s^{-2}, where $a = 0.05 - 0.0002t$.

i Find A's velocity when $t = 200$ and when $t = 500$. [4]

B moves with constant acceleration for the first 200 s and has the same velocity as A when $t = 200$. B moves with constant retardation from $t = 200$ to $t = 500$ and has the same velocity as A when $t = 500$.

ii Find the distance between A and B when $t = 500$. [6]

Cambridge International AS & A Level Mathematics 9709 Paper 41 Q6 June 2015

11 A vehicle is moving in a straight line. The velocity v m s^{-1} at time t s after the vehicle starts is given by $v = A(t - 0.05t^2)$ for $0 \leq t \leq 15$ and $v = \frac{B}{t^2}$ for $t \geq 15$, where A and B are constants. The distance travelled by the vehicle between $t = 0$ and $t = 15$ is 225 m.

i Find the value of A and show that $B = 3375$. [5]

ii Find an expression in terms of t for the total distance travelled by the vehicle when $t \geq 15$. [3]

iii Find the speed of the vehicle when it has travelled a total distance of 315 m. [3]

Cambridge International AS & A Level Mathematics 9709 Paper 42 Q7 June 2010

12 Two walkers, P and Q, travel along a straight track ABC. Both walkers start from point A at time $t = 0$ s and pass through point B at time $t = 10$ s. They both finish at point C. P starts from point A with speed 2 m s^{-1} and accelerates with constant acceleration 0.1 m s^{-2} until reaching point B.

a Show that the distance from A to B is 25 m. [3]

b Find the speed of P on reaching point B. [2]

Q starts from point A and moves with speed v_1 m s^{-1}, given by $v_1 = 0.003t^2 + 0.06t + k$. When Q passes through point B both walkers have the same speed.

c Find the value of the constant k. [3]

P moves from point B to point C with speed v_2 m s^{-1}, given by $v_2 = 4 - 0.1t$, and comes to rest as C is reached.

d Show that the distance from A to C is 70 m. [4]

Q moves from point B to point C with speed v_3 m s^{-1}, given by $v_3 = 0.4t - 0.01t^2$.

e Show that Q reaches point C first. [3]

CROSS-TOPIC REVIEW EXERCISE 2

1. Two particles, A and B, are attached to the ends of a light inextensible string, which passes over a smooth pulley. Particle A has mass $4\,\text{kg}$ and B has mass $6\,\text{kg}$. The system is released from rest and the particles move vertically.

   a Find the tension in the string and the upward acceleration of particle A. [3]

   b Find the magnitude of the resultant force exerted on the pulley by the string. [1]

2. A particle of mass $3\,\text{kg}$ is at rest on a slope that is at an angle of $27°$ to the horizontal. It is held in limiting equilibrium by a force of $5\,\text{N}$, which acts at an angle of $10°$ to the slope, as shown. Determine in which direction the particle is on the point of slipping and find the coefficient of friction between the particle and the slope. [5]

3. A particle P, with mass $3\,\text{kg}$, and a particle Q, with mass $5\,\text{kg}$, are attached to the ends of a light inextensible string. P is held at rest on a horizontal table and the coefficient of friction between P and the table is 0.4. The string passes over a smooth pulley at the end of the table $0.8\,\text{m}$ from P and Q hangs vertically down, as shown.

   The particles are then released from rest. Find the time until particle P hits the pulley. [5]

4. A toy train engine has mass $4\,\text{kg}$ and pulls a carriage of mass $6\,\text{kg}$ along a horizontal stretch of track by means of a horizontal tow-bar. The brakes in the engine cause a deceleration of $2\,\text{m s}^{-2}$. There is air resistance of $4\,\text{N}$ on the engine and of $10\,\text{N}$ on the carriage and no other frictional forces. Find the braking force from the engine and the force in the tow-bar, stating whether it is a tension or compression. [4]

5. A particle, P, starts from a point O and moves in a straight line with velocity $v\,\text{m s}^{-1}$, given by:

   $v = k$ for $0 < t \leqslant 1$

   $v = 3t^{\frac{1}{2}} + \dfrac{24}{t^2}$ for $1 \leqslant t \leqslant 5$

   where t is the time, in seconds, after leaving O.

   a Find the minimum velocity for $1 \leqslant t \leqslant 5$. [4]

   b Find the displacement from O when P reaches the minimum velocity. [5]

6. A particle P, of mass $4\,\text{kg}$, is projected from a point A up a slope with speed $5\,\text{m s}^{-1}$. The slope is at an angle of $25°$ to the horizontal and the coefficient of friction between the slope and the particle is 0.4.

   a Find the distance P travels up the slope before coming to rest. [4]

   b Find the time taken for P to return to A. [5]

7. Two particles move along the same straight line. Particle P has velocity $v\,\text{m s}^{-1}$, given by $v = 12t + t^3 - 0.3t^5$, where t is the time in s, and is at a point O at $t = 0$. Particle Q has displacement s from O at time t, given by $s = 74.25 - 0.05t^6$.

   a Find the displacement of P when it is moving at maximum velocity. [6]

   b The particles collide at time T. Find the value of T. [2]

8 Two particles, A and B, are attached to the ends of a light inextensible string, which passes over a smooth pulley. Particle A has mass 8 kg and B has mass 5 kg. Both particles are held 1.2 m above the ground. The system is released from rest and the particles move vertically.

  a When particle A hits the ground, it does not bounce. Find the maximum height reached by particle B. [5]

  b When particle A hits the ground, the string is cut. Find the total time from being released from rest until B hits the ground. [3]

9 A particle, P, moves on a straight track. It has displacement s m from a point, O, at time t s, given by $s = 0.12t + 10 - 0.01t^3$ for $t > 0$.

  a Find the time when the particle is first stationary. [2]

  b Find the total distance travelled in the first 10 s. [2]

  Particle Q moves on a track parallel to particle P. The acceleration, a m s^{-2}, of Q is given by $a = 0.4 - 0.06t$. Both particles come to rest alongside each other.

  c Find the displacement of Q from O after 10 s. [5]

10 Two particles of masses 5 kg and 10 kg are connected by a light inextensible string that passes over a fixed smooth pulley. The 5 kg particle is on a rough fixed slope which is at an angle of α to the horizontal, where $\tan \alpha = \dfrac{3}{4}$. The 10 kg particle hangs below the pulley (see diagram). The coefficient of friction between the slope and the 5 kg particle is $\dfrac{1}{2}$. The particles are released from rest. Find the acceleration of the particles and the tension in the string. [7]

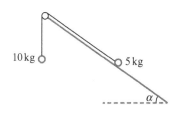

Cambridge International AS & A Level Mathematics 9709 Paper 41 Q5 June 2016

11 A particle of mass 0.1 kg is released from rest on a rough plane inclined at 20° to the horizontal. It is given that, 5 seconds after release, the particle has a speed of 2 m s^{-1}.

  i Find the acceleration of the particle and hence show that the magnitude of the frictional force acting on the particle is 0.302 N, correct to 3 significant figures. [3]

  ii Find the coefficient of friction between the particle and the plane. [2]

Cambridge International AS & A Level Mathematics 9709 Paper 41 Q2 November 2016

12 A particle P moves in a straight line. It starts at a point O on the line and at time t s after leaving O it has a velocity v m s^{-1}, where $v = 6t^2 - 30t + 24$.

  i Find the set of values of t for which the acceleration of the particle is negative. [2]

  ii Find the distance between the two positions at which P is at instantaneous rest. [4]

  iii Find the two positive values of t at which P passes through O. [3]

Cambridge International AS & A Level Mathematics 9709 Paper 41 Q6 June 2016

13 Particles P and Q are attached to opposite ends of a light inextensible string which passes over a fixed smooth pulley. The system is released from rest with the string taut, with its straight parts vertical, and with both particles at a height of 2 m above horizontal ground. P moves vertically downwards and does not rebound when it hits the ground. At the instant that P hits the ground, Q is at the point X, from where it continues to move vertically upwards without reaching the pulley. Given that P has mass 0.9 kg and that the tension in the string is 7.2 N while P is moving, find the total distance travelled by Q from the instant it first reaches X until it returns to X. [6]

Cambridge International AS & A Level Mathematics 9709 Paper 43 Q3 November 2011

Chapter 7
Momentum

In this chapter you will learn how to:

- calculate the momentum of a moving body or a system of bodies
- use the principle of conservation of momentum to solve problems involving the direct impact of two bodies that separate after impact
- use the principle of conservation of momentum to solve problems involving the direct impact of two bodies that coalesce on impact.

PREREQUISITE KNOWLEDGE

Where it comes from	What you should be able to do	Check your skills
Chapter 1	Calculate velocity when the acceleration is constant.	1 A car is travelling at $15\,\text{m s}^{-1}$ when the brakes are applied. It takes 6 s for the car to come to rest. Assume that the braking force is constant (and hence the acceleration is constant, but negative). **a** Show that car travels 45 m under braking before coming to rest. **b** Calculate speed of the car when it has been braking for 3 s. **c** Calculate the speed of the car when it has travelled 22.5 m under braking.
Chapter 6	Calculate velocity using calculus.	2 A car is travelling at $15\,\text{m s}^{-1}$ when the brakes are applied. It takes 6 s for the car to come to rest. Assume that the acceleration under braking is given by $\dfrac{5t(t-6)}{12}$ where t is the time from when braking starts. **a** Calculate the speed of the car when it has been braking for 3 s. **b** Find the speed of the car when it has travelled 22.5 m under braking.

What is momentum?

The word momentum is used in everyday language to describe the impetus gained:

> *Feminism gained momentum in the early 20th century.*

> *The fundraising campaign needs to gain momentum if it is to reach its goal.*

In mechanics, momentum measures the impetus possessed by a moving object. By considering the transfer of momentum between objects you can calculate what happens when objects interact.

You may have pushed a supermarket trolley. Why is it easier to start the trolley moving when it is empty than when it is full of shopping? An empty trolley has less mass than a full trolley, so the same amount of 'push' will get an empty trolley moving much faster than a full trolley.

Chapter 7: Momentum

You explain this in mechanics by using momentum.

7.1 Momentum

 DID YOU KNOW?

 The philosopher René Descartes (1596–1650) introduced the concept of momentum. Descartes' built on ideas first written down by Jean Buridan (1295–1363) who defined the 'amount of motion' as the product of the mass of a body and its speed. Using these ideas, Descartes formulated his three laws of motion, which then became the basis for Newton's laws of motion.

KEY POINT 7.1

A body of mass m kg moving with speed v m s^{-1} has momentum given by mv.

Momentum is a vector quantity, having the same direction as the velocity. For one-dimensional motion along a line you only need to work out whether the momentum is positive or negative. The units of momentum are N s.

EXPLORE 7.1

In the Système Internationale (SI) system of units there are seven basic units of measurement. These are the metre (length), kilogram (mass), second (time), ampere (electrical current), kelvin (thermodynamic temperature), mole (amount of substance) and candela (luminous intensity). Use the SI system of units to explain why momentum is measured in N s.

WORKED EXAMPLE 7.1

Find the momentum of a body of mass 3 kg moving at 5 m s^{-1}.

Answer

Momentum $= mv = 3 \times 5$ .... Substitute the values for m and v into the formula for momentum.

$= 15$ N s .... Remember to give units.

Cambridge International AS & A Level Mathematics: Mechanics

WORKED EXAMPLE 7.2

A ball of mass 50 g hits the ground with speed $10\,\text{m s}^{-1}$ and rebounds with speed $6\,\text{m s}^{-1}$. Find the change in momentum that occurs in the bounce.

Answer

$50\,\text{g} = 0.050\,\text{kg}$	Convert the mass to kg.
Momentum before $= 0.050 \times 10$ $= 0.5\,\text{N s}$	Calculate the momentum just before the bounce.
Momentum after $= 0.050 \times -6$ $= -0.3\,\text{N s}$	The direction has reversed so the sign changes.
So change in momentum $= -0.3 - 0.5$ $= -0.8\,\text{N s}$	If we use down as positive there is a loss in momentum of $0.8\,\text{N s}$.
	If we use up as positive there is a gain in momentum of $0.8\,\text{N s}$.

EXERCISE 7A

1. Find the momentum of a body of mass 10 kg moving at $8\,\text{m s}^{-1}$.

2. Find the momentum of a car of mass 1500 kg moving at $22\,\text{m s}^{-1}$.

3. Find the momentum of a tennis ball of mass 57 g moving at $180\,\text{km h}^{-1}$.

4. A model car has mass 40 g. It slows from $2.2\,\text{m s}^{-1}$ to $0.8\,\text{m s}^{-1}$. Find the decrease in its momentum.

5. A rock of mass 4 kg is thrown upwards with an initial speed of $3\,\text{m s}^{-1}$. It is travelling at $6\,\text{m s}^{-1}$ just before it lands. Find the change in its momentum.

6. A girl of mass 35 kg jumps from a rock onto the beach below. Her initial vertical speed is $0\,\text{m s}^{-1}$ and she falls 2.45 m under gravity.

    a Find the speed of the girl when she lands on the beach.

    b Find the downward momentum of the girl just before she lands on the beach.

7. A book of mass 2 kg falls from a window ledge and drops 1.8 m to the ground. It falls freely under gravity.

    a Find the speed of the book just before it hits the ground.

    b Find the downward momentum of the book just before it hits the ground.

8. A ball of mass 0.2 kg falls 1.25 m vertically downwards to the ground, starting from rest. It hits the ground and rebounds. The downwards momentum of the ball changes by $1.6\,\text{N s}$ in the bounce.

    a What height does the ball reach after this bounce?

    b By considering the modelling assumptions, explain why the height might be less than this.

9 A ball bearing of mass 25 g is thrown vertically upwards, and is caught on the way back down. The ball bearing has an initial speed of $3\,\text{m s}^{-1}$ upwards and is travelling at $2\,\text{m s}^{-1}$ when it is caught. Find the change in its momentum.

10 A hockey ball of mass 0.2 kg is hit so that it has an initial speed of $8\,\text{m s}^{-1}$. The ball travels in a horizontal straight line with acceleration $a\,\text{m s}^{-2}$ given by $a = -0.5 - kt$ where t is the time in seconds, measured from when the ball was hit. After 2 s the ball has travelled $\frac{41}{3}$ m. It is then intercepted by a player from the other team. This player hits the ball so that its direction of travel is reversed and its speed is now $5\,\text{m s}^{-1}$. Show that when the ball is hit by the second player its momentum changes in magnitude by 2 N s.

11 Particle A of mass 5 kg is moving at a speed of $2\,\text{m s}^{-1}$ when it hits a stationary particle, B, of mass 2 kg. After the impact, particle A has speed $0\,\text{m s}^{-1}$ and particle B has speed $v\,\text{m s}^{-1}$. The loss in momentum for particle A equals the gain in momentum for particle B. Find the value of v.

12 A man strikes a snooker ball so that it travels horizontally across a snooker table and makes a direct hit against the end cushion of the table. The ball rebounds from the cushion and travels to the other end of the table where it rebounds from the cushion at that end. The ball finishes at exactly the same point at which it started. The distance between the two cushions is 3.5 m and the initial speed of the ball is $10\,\text{m s}^{-1}$. The ball is slowed by friction, resulting in a constant deceleration of $1\,\text{m s}^{-2}$. At each rebound the direction of the ball is reversed and the magnitude of the momentum after the rebound is 50% of the magnitude of the momentum before. Work out the distance that the ball travels before it reaches the first cushion.

7.2 Collisions and conservation of momentum

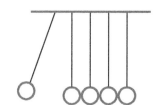

During an **impact** when two bodies collide, there is a transfer of momentum between them. Some momentum will be transferred from the first to the second and some momentum will be transferred from the second to the first. In this chapter you will only consider one-dimensional impacts between bodies moving in the same straight line, both before and after the impact.

Newton's cradle, shown in the diagram, is a popular toy. The first ball is released and transfers momentum to the second, which in turn transfers momentum to the third, and so on until the last ball swings up. The last ball then swings back down again and the momentum is transferred back to the first ball.

In a perfect Newton's cradle, each ball comes to rest after it has hit the next one, so it looks as if the first ball has caused the last one to move without the intermediate balls moving at all.

When a hammer is used to hit a nail, momentum is transferred from the hammer to the nail, causing the nail to move (although resistance forces mean that the nail will not move very far). Momentum is also transferred in the opposite direction, causing the hammer to rebound.

Impacts happen instantaneously, so you do not need to think about external forces, such as friction, when considering the impact. The change in momentum is caused by the normal contact forces between the two objects involved.

One-dimensional instantaneous impacts happen, for example, when a snooker ball, with no spin, hits a stationary ball and causes it to move along the same line as the original motion.

The contact forces between the two objects involved in the impact are equal and opposite, so the momentum transferred from the first object to the second is equal and opposite to the momentum transferred from the second object to the first.

This means that the total momentum before the impact will always be the same as the total momentum after the impact. The total momentum is unchanged: momentum is conserved in an impact.

MODELLING ASSUMPTIONS

In reality a snooker player would not usually want a direct, one-dimensional impact and would probably prefer to use an oblique, two-dimensional impact, where the motion is not all in the same straight line

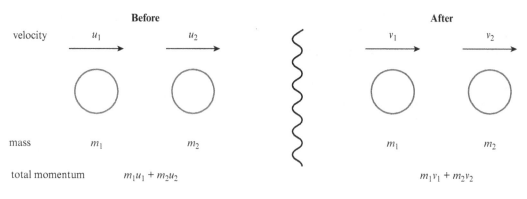

KEY POINT 7.2

Momentum is conserved in impacts. The total momentum is constant.

WORKED EXAMPLE 7.3

Two ball bearings are moving directly towards one another. The first ball bearing has mass 20 g and is moving at $3\,\text{m s}^{-1}$. The second ball bearing has mass 25 g and is moving at $1\,\text{m s}^{-1}$. After the collision the first ball bearing is stationary. What is the speed of the second ball bearing after the collision?

Answer

before		after	
$3\,\text{m s}^{-1}$	$1\,\text{m s}^{-1}$	$0\,\text{m s}^{-1}$	v
→	←	→	→
○	○	○	○
0.020 kg	0.025 kg	0.020 kg	0.025 kg

Draw a diagram to summarise the information.

Total momentum before collision

$= (0.020 \times 3) + (0.025 \times -1)$

$= 0.035\,\text{N s}$

Remember that momentum is a vector quantity.

$1\,\text{m s}^{-1} \leftarrow$ is the same as $-1\,\text{m s}^{-1} \rightarrow$

Total momentum after collision
$$= 0 + 0.025v$$
$$= 0.025v \text{ N s}$$
$$0.035 = 0.025v$$ .... Momentum is conserved.
$$v = 1.4 \text{ m s}^{-1}$$

Sometimes, instead of moving apart after an impact, the objects may **coalesce**. This means that they collide and then move off together as a single object. The objects can be thought of as having merged into a single object with a mass equal to the sum of the individual masses.

Examples of coalescence include a railway truck being pushed up to an engine and coupling with it, a person jumping onto a moving vehicle or two ice skaters meeting up and holding hands to continue as one.

The opposite of coalescence is called an **explosion**. This would happen, for example, when the engine and truck become decoupled, when the person jumps off the moving vehicle or when the ice skaters stop holding hands and drift apart.

WORKED EXAMPLE 7.4

A girl is sitting on a sledge. The girl and the sledge have a combined mass of 50 kg. When the girl and the sledge are moving at 2 m s^{-1} her sister standing in front of the sledge throws a snowball at the sledge. The snowball has mass 0.2 kg and is travelling at 10.55 m s^{-1} when it hits the sledge, head on. The snowball, the girl and the sledge continue together. Assuming that the total momentum is unchanged, find the new speed of the sledge.

Answer

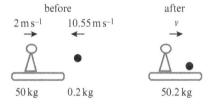

Draw a diagram to summarise the information.

Total momentum before $= (50 \times 2) + 0.2 \times (-10.55)$
$$= 97.89 \text{ N s}$$

The snowball is thrown in the opposite direction to the motion of the sledge.

Total momentum after $= 50.2v \text{ N s}$

The total mass of the girl, sledge and snowball is $50 + 0.2 = 50.2 \text{ kg}$.

$$97.89 = 50.2v$$ .... Momentum is conserved.
$$v = 1.95 \text{ m s}^{-1}$$

WORKED EXAMPLE 7.5

A block of mass 200 g moving at $4\,\text{m s}^{-1}$ makes a direct collision with a larger block of mass 500 g moving at $1\,\text{m s}^{-1}$. On impact the blocks coalesce.

  a Find the speed of the blocks after the collision if the blocks were initially moving towards one another.

  b Find the speed of the blocks after the collision if the blocks were initially moving in the same direction.

Answer

a

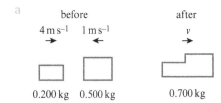

Total momentum before $= (0.200 \times 4) + (0.500 \times -1)$ $= 0.3\,\text{N s}$ Total momentum after $= 0.700v\,\text{N s}$	Momentum is a vector quantity.
$0.3 = 0.700v$ $v = \dfrac{3}{7}\,\text{m s}^{-1}$ $= 0.429\,\text{m s}^{-1}$ (to 3 s.f.)	Momentum is conserved.

b

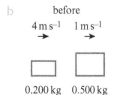

	The faster block must catch up with the slower block.
Total momentum before $= (0.200 \times 4) + (0.500 \times 1)$ $= 1.3\,\text{N s}$ Total momentum after $= 0.700v\,\text{N s}$	
$1.3 = 0.700v$ $v = \dfrac{13}{7}\,\text{m s}^{-1}$ $= 1.86\,\text{m s}^{-1}$ (to 3 s.f.)	Momentum is conserved.

MODELLING ASSUMPTIONS

By modelling objects as particles, you are ignoring the possibility of an oblique contact. This can happen if objects collide so that the contact is not in the line of motion and instead they bounce off each other at different angles. You will assume this is not the case.

There is also a possibility that objects travelling along a surface wobble slightly or if the objects are a different size or shape some of the momentum may cause one of the objects to lift off the surface. The effect of this is normally quite small, but can be significant in games where precision is required.

EXERCISE 7B

1 Chris and his son are skating on an ice rink. Chris skates in a straight line at a speed of $3\,\mathrm{m\,s^{-1}}$ towards his son, who is stationary on the ice. When they meet Chris lifts his son up and they continue together at a speed of $2\,\mathrm{m\,s^{-1}}$, still travelling in the same straight line. Chris has mass $80\,\mathrm{kg}$. Find the mass of his son.

2 A ball of mass $0.04\,\mathrm{kg}$ is moving at a speed of $3\,\mathrm{m\,s^{-1}}$ when it hits a stationary ball of mass $0.06\,\mathrm{kg}$. After the impact the first ball is stationary. Find the speed of the second ball.

3 A box of mass $25\,\mathrm{kg}$ slides down a slope until it has reached a speed of $5\,\mathrm{m\,s^{-1}}$ It then travels at $5\,\mathrm{m\,s^{-1}}$ horizontally across a smooth floor until it bumps into a stationary crate. Immediately after the impact the box reverses its direction and travels at $0.5\,\mathrm{m\,s^{-1}}$. The crate starts to travel at $2.75\,\mathrm{m\,s^{-1}}$. Find the mass of the crate.

PS 4 Two snooker balls are travelling towards one another in a straight line when they make a direct impact. Before the impact the first ball had speed $12\,\mathrm{m\,s^{-1}}$ and the second ball had speed $8\,\mathrm{m\,s^{-1}}$. After the impact both balls have reversed their direction and each has speed $10\,\mathrm{m\,s^{-1}}$. It is claimed that the balls are not both real snooker balls because they have different masses. Find the ratio of the masses of the balls.

5 Particles A, B and C, of masses $0.01\,\mathrm{kg}$, $0.06\,\mathrm{kg}$ and $0.12\,\mathrm{kg}$ respectively, are at rest in a straight line on a smooth horizontal surface, with B between A and C. A is given an initial velocity of $4\,\mathrm{m\,s^{-1}}$ towards B. After this impact A rebounds with velocity $2\,\mathrm{m\,s^{-1}}$ and B goes on to hit C. After the second impact B comes to rest. Find the speed of C after the second impact.

6 Three balls, A, B and C, of masses $4\,\mathrm{kg}$, $1\,\mathrm{kg}$ and $2\,\mathrm{kg}$, respectively, are at rest in a straight line on a smooth horizontal surface, with B between A and C. A is given an initial velocity of $5\,\mathrm{m\,s^{-1}}$ towards B. When A hits B they coalesce and continue as a single object, D, until they collide with C. After this collision C has velocity $5\,\mathrm{m\,s^{-1}}$. Work out the final velocity of D.

M 7 Jayne is performing in a show on ice. She is pushed onto the ice while sitting on a chair. The chair slides across the ice and Jayne then stands up and moves away from the chair. Jayne has speed $4\,\mathrm{m\,s^{-1}}$ when she is sitting on the chair and speed $5\,\mathrm{m\,s^{-1}}$ when she moves away from the chair. Jayne has mass $60\,\mathrm{kg}$ and the chair has mass $6\,\mathrm{kg}$.

  a Find the velocity of the chair as Jayne moves away from it.

  b What modelling assumptions have you made?

8 A bean bag of mass 100 g is thrown at $5\,\text{m s}^{-1}$ at a stationary target. The bean bag sticks to the target and they move off together at $0.1\,\text{m s}^{-1}$. Find the mass of the target.

9 Mariam is moving on a sledge at $2\,\text{m s}^{-1}$. The combined mass of Mariam and the sledge is 40 kg. Sarah, who has mass 60 kg, runs up behind the sledge and jumps onto it. The sledge continues in the same straight line with speed $2.3\,\text{m s}^{-1}$.

   a Find Sarah's speed just before she lands on the sledge.

   b What assumption have you made regarding Sarah's velocity?

10 A simplified model of the launch of a space shuttle is as follows. The shuttle is attached to two rocket boosters each of which contains a fuel tank. The launch is vertical and in a straight line. The initial total mass is 1 million kg. The mass of the shuttle is 60 000 kg, the mass of each rocket booster is 20 000 kg and the mass of the fuel in each rocket booster is 450 000 kg. The rocket boosters accelerate the shuttle (and themselves) to a speed of $1500\,\text{m s}^{-1}$. At this time all the fuel in the rocket boosters has been used up and the rocket boosters are detached. The rocket boosters have speed $0\,\text{m s}^{-1}$ immediately after they are detached.

   a Use conservation of momentum to show that the speed of the shuttle immediately after the rocket boosters are detached is $2500\,\text{m s}^{-1}$.

Suppose that instead just the first rocket booster is used initially to accelerate the shuttle (with both rocket boosters) to a speed of $500\,\text{m s}^{-1}$. At this time all the fuel in the first rocket booster has been used up and it is detached (with speed $0\,\text{m s}^{-1}$). The second rocket booster is still full of fuel.

   b Show that the speed of the shuttle (with the remaining rocket booster) is $518.9\,\text{m s}^{-1}$ immediately after the first rocket booster is detached.

The second rocket booster is then used to accelerate the shuttle (and itself). When all the fuel in the second rocket booster has been used up it is detached (with speed $0\,\text{m s}^{-1}$). After the second rocket booster has been detached, the speed of the shuttle is $2500\,\text{m s}^{-1}$.

   c Find the speed of the shuttle (and rocket booster) just before the second rocket booster was detached.

11 A car is towing a caravan at $v\,\text{m s}^{-1}$ in a straight line along a horizontal road. The mass of the car is m kg and the mass of the caravan is km kg. The caravan becomes detached from the car. Immediately after the separation the car has speed $\alpha v\,\text{m s}^{-1}$ and the caravan has speed $0.5v\,\text{m s}^{-1}$ in the same direction as the car.

   a Show that if $\alpha = 1.3$ then $k = 0.6$.

   b Find an expression for k in terms of a general value of α.

12 A particle of mass 0.3 kg is travelling at speed $2\,\text{m s}^{-1}$ when it collides with a particle of mass 0.5 kg travelling at speed $1\,\text{m s}^{-1}$. After the impact the first particle has speed $v\,\text{m s}^{-1}$ and the second particle has speed $(v+1)\,\text{m s}^{-1}$.

   a By considering the directions in which the particles could be moving before and after the impact, find the possible values for the speed of the first particle after the impact.

   You are given that v is the smallest of these possible speeds.

   b State whether the particles were travelling in the same direction or in opposite directions before the impact.

EXPLORE 7.2

Five small balls are placed in a line on a smooth table with 1 m between each ball and the next. The fifth ball is at the edge of the table. The first ball has mass 50 g, the second has mass 40 g, the third has mass 30 g, the fourth has mass 20 g and the fifth has mass 10 g. Initially the balls are all stationary. The first ball is then fired to hit the second ball with speed 1 m s^{-1}. In this collision, half the momentum of the first ball is transferred to the second ball. It is claimed that in every impact, half the momentum of the faster ball is transferred to the slower ball. Investigate what happens and how long it takes until the fifth ball falls from the table.

Checklist of learning and understanding

- A body of mass m kg moving with speed v m s^{-1} has momentum given by mv.
- Momentum is conserved in impacts. The total momentum is constant.
- $m_1 u_1 + m_2 u_2 = m_1 v_1 + m_2 v_2$

END-OF-CHAPTER REVIEW EXERCISE 7

1. Particle A moves across a smooth horizontal surface in a straight line. Particle A has mass 4 kg and speed $3\,\text{m s}^{-1}$. Particle B, which has mass 6 kg, is at rest on the surface. Particle A collides with particle B. After the collision A is at rest and B moves away from A with speed $u\,\text{m s}^{-1}$. Find the value of u. [3]

2. Two particles, A and B, have masses of 3 kg and 2 kg, respectively. They are moving along a straight horizontal line towards each other. Each particle is moving with a speed of $4\,\text{m s}^{-1}$ when they collide.

   The particles coalesce to form a single particle. Find the speed of the combined particle. [3]

   $A\,(3\,\text{kg})$ $B\,(2\,\text{kg})$

   $4\,\text{m s}^{-1}$ $4\,\text{m s}^{-1}$

3. A train consists of a locomotive, of 40 000 kg, pulling four coaches, each of mass 50 000 kg. The train is travelling at $5\,\text{m s}^{-1}$ along a straight horizontal line when the coupling between the second and third coaches breaks. The locomotive and the first two coaches continue at $7\,\text{m s}^{-1}$. The rear two coaches then decelerate at a constant rate to come to rest after travelling 100 m. Work out how long it takes from when the coupling breaks to when the rear two coaches come to rest. [4]

4. Particle A has mass 4 kg. It moves with speed $3\,\text{m s}^{-1}$ in a straight line on a smooth horizontal surface. Particle B has mass 6 kg and is at rest on the surface. Particle A collides with particle B. After the collision, A and B move away from each other with speeds $v\,\text{m s}^{-1}$ and $4v\,\text{m s}^{-1}$, as shown in the diagram.

   Find the value of v. [4]

5. Two balls are travelling towards one another along the x-axis. The first ball has mass 2 kg and is travelling at $3\,\text{m s}^{-1}$ in the positive x-direction. The second ball has mass 5 kg is travelling at $1\,\text{m s}^{-1}$ in the negative x-direction. The balls collide and after the collision the balls are travelling at the same speed but in opposite directions. Work out the speed of the balls after the collision. [4]

6. Two particles, A and B, are moving in a straight line on a smooth horizontal surface. A has mass m kg and is moving with velocity $5\,\text{m s}^{-1}$, B has mass 0.2 kg and is moving with velocity $2\,\text{m s}^{-1}$.

   a Find, in terms of m, an expression for the total momentum of A and B. [1]

   Particle A collides with particle B and they coalesce to form a single particle, C. Particle C has velocity $3\,\text{m s}^{-1}$.

   b Find the value of m. [3]

7. Two particles, A and B, have masses of 3 kg and 2 kg, respectively. They are moving along a straight horizontal line towards each other. Each particle is moving with a speed of $4\,\text{m s}^{-1}$ when they collide.

   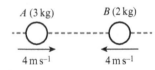

   After the collision, particle A moves in the same direction as before the collision but with speed $0.4\,\text{m s}^{-1}$. Find the speed of B after the collision. [4]

8. Ball X has mass 0.03 kg. It falls vertically from rest from a window that is 30 m above the ground. Ball Y has mass 0.01 kg. At the same time that ball X starts to fall, ball Y is projected vertically upwards from ground level directly towards ball X. The initial speed of ball Y is $20\,\text{m s}^{-1}$ vertically upwards.

   a Find the downward momentum of each ball just before they meet. [3]

   The balls coalesce and the combined object falls to the ground.

   b Show that the combined object reaches the ground 2.68 seconds after ball X started to fall. [5]

9 Three balls, A, B and C, are in that order in a straight line on a smooth horizontal surface. A has mass $0.4\,\text{kg}$ and is moving at $4\,\text{m s}^{-1}$ towards B. B has mass $m\,\text{kg}$ and is stationary. C has mass $0.25\,\text{kg}$ and is moving at $0.8\,\text{m s}^{-1}$ away from B. A hits B and then B hits C. There are no further impacts. A and C now each have a speed of $1\,\text{m s}^{-1}$ and are both moving in directions away from B. Find the range of possible values of m. [8]

10 A ball of mass $0.6\,\text{kg}$ is dropped from a height of $1.8\,\text{m}$ onto a solid floor. Each time the ball bounces on the floor it loses 10% of its speed.

   a Work out how much momentum was absorbed by the floor in the first bounce. [3]

   b Show that the ball first fails to reach a height of $1\,\text{m}$ after the third bounce. [4]

   c What modelling assumptions have you made? [1]

11 Ball X has mass $30\,\text{g}$ and is moving at $0.51\,\text{m s}^{-1}$. The direction in which X is travelling is taken as the positive direction. Ball Y has mass $60\,\text{g}$ and is stationary. Ball X collides with ball Y and, after the impact, ball X moves at $0.01\,\text{m s}^{-1}$ in the positive direction. Ball Y then hits a wall and rebounds with half the speed with which it hit the wall.

   a Work out how much momentum was absorbed by the wall. [5]

   After rebounding from the wall, ball Y goes on to hit ball X.

   b Explain why ball X must be travelling in the negative direction after being hit by ball Y. [2]

   After this impact ball X has speed $0.15\,\text{m s}^{-1}$.

   c Find the final velocity of ball Y. [3]

12 Balls X, Y and Z lie at rest on a smooth horizontal surface, with Y between X and Z. Balls X and Z each have mass $2\,\text{kg}$ and ball Y has mass $1\,\text{kg}$. Ball X is given a velocity of $1\,\text{m s}^{-1}$ towards ball Y. Balls X and Y collide. After this collision the speed of ball Y is three times the speed of ball X. Ball Y goes on to collide with ball Z. After this collision the speed of ball Y is the same as the speed of ball X, and the speed of ball Z is twice the speed of ball Y. Finally ball Y collides with ball X again. After this collision the speed of ball Y is twice the speed of ball X, and the speed of ball Z is four times the speed of ball Y. Show that the balls are now all travelling in the same direction and that no further collisions occur. [10]

Chapter 8
Work and energy

In this chapter you will learn how to:

- calculate the work done by a force in moving a body
- calculate the kinetic energy and gravitational potential energy of a body.

Chapter 8: Work and energy

PREREQUISITE KNOWLEDGE

Where it comes from	What you should be able to do	Check your skills
Chapter 3	Resolve forces.	1 A block of mass 4 kg is at rest on a slope that is inclined at 30° to the horizontal. A force parallel to the slope prevents the block from moving. a Find the component of the block's weight down the slope. b Find the normal reaction that the slope exerts on the block.
Chapter 4	Calculate frictional resistance.	2 A block of mass 4 kg is sliding down a slope. The coefficient of friction between the slope and the block is $\frac{1}{10}\sqrt{3}$. The normal reaction that the slope exerts on the body is $20\sqrt{3}$ N. Find the frictional force.
Chapter 2	Use Newton's second law.	3 A block of mass 4 kg is sliding down a slope. The component of the weight down the slope is 20 N and the frictional force up the slope is 2 N. a Find the resultant force on the block. b Find the acceleration of the block down the slope.
Chapter 1	Use the equations of constant acceleration.	4 A block is initially at rest on a slope. It slides down the slope with constant acceleration $4.5 \, \text{m s}^{-2}$ down the slope. How far does the block slide in the first 0.5 s?

How are work and energy used in mechanics?

The terms work and energy are used in everyday life, but what do these terms mean when we use them in mechanics and how are they connected?

In everyday life, a student who has been studying hard for 2 hours would say that they have been doing work, as would a gardener who has been working in a garden or an athlete who has been training. Each of these people has spent time doing an activity and has used energy in the process.

This energy comes from the food that the people have eaten. The gardener and the athlete have used energy to create movement, and the student has used energy to create 'brainpower'.

The phrase 'put more energy into it' is used to mean put more effort into a task, or apply more force. Energy comes in many forms and can be changed from one form to another. A person who eats a meal takes in chemical energy, which might then be converted into movement or, if it is a cold day, used to warm the person up.

In this chapter we will show that when a force moves a body, it does work and causes a change in the kinetic energy of the body. In Chapter 9 we will further investigate this relationship between work and energy.

8.1 Work done by a force

In mechanics the word **work** means something more than just making an effort. It has a very specific meaning that refers to how energy changes when a force moves an object.

Mechanical work is done by a force when that force causes an object to move. For mechanical work to happen, we need a force that causes motion *and* we need motion to occur.

A weightlifter does work in lifting a weight because a force acts (the tension in the arm of the weightlifter) to cause motion (the weight is raised vertically).

However, no mechanical work is done when the weightlifter holds the weight stationary above their head, because there is no motion (although clearly it requires a lot of effort to stop the weight from falling).

We start by considering the work done by a force acting in the direction of motion; for example, a horizontal force pushing a box across a horizontal floor.

If the force doubles then the work done by the force doubles. The work done would also double if the force was unchanged but the object moved twice as far.

The **line of action** of a force has the same direction as the force and includes the point of application of the force.

KEY POINT 8.1

When a force of magnitude F N moves a body a distance d m, along the line of action of the force, the work done by the force is:

$$W = Fd$$

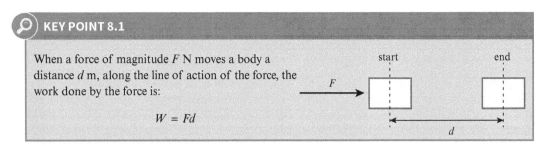

FAST FORWARD

Later in this section, we will consider what it means for work done to be negative.

Note that the distance moved has been represented by d here. When the motion is in a straight line and in a constant direction, the distance moved will be the same as the displacement, s, and then the work done by the force is given by $W = Fs$.

Work is a scalar quantity; it can be positive or negative, but otherwise has no direction.

The work done by a force has units N m, but it is more usual to use joules (J) to measure work done. One joule is the amount of work done by a force of 1 newton in moving an object a distance of 1 m, along the line of action of the force.

$$1 J = 1 N m$$

DID YOU KNOW?

James Prescott Joule (1818–1889) studied the nature of heat and discovered its relationship to mechanical work. This led to the development of the first law of thermodynamics.

Chapter 8: Work and energy

WORKED EXAMPLE 8.1

A boy uses a constant force of 250 N to push a box 4 m across a floor. Find the work done by the force.

Answer

Work done = $Fd = 250 \times 4$ .... Substitute the values for F and d into the formula for work done.

$= 1000$ J .... Remember to give units.

WORKED EXAMPLE 8.2

A girl holds a mass of 20 kg above her head. Find the work done by the girl.

Answer

The mass does not move, so no work is done. .... This shows how mechanical work differs from the everyday use of the word work.

Work done = 0 J .... Work is done in raising the mass but no mechanical work is done in holding it steady, despite how it might feel!

WORKED EXAMPLE 8.3

A ball of mass 0.05 kg falls a distance of 1.5 m. Find the work done by the weight.

Answer

Weight = 0.05×10 .... The ball falls because of gravity, so the force

$= 0.5$ N that is causing the ball to fall is its weight.

Work done = 0.5×1.5 .... This is the work done by the weight.

$= 0.75$ J

The work done by the weight of an object is usually referred to as the **work done by gravity**. If other forces act and the direction of motion is upwards, we describe this as the **work done against gravity**.

Sometimes the direction of motion can be in a different direction to the line of action of the force that is causing the motion. This can occur when there are other forces acting. For example, when a force at an angle to the horizontal pushes or pulls a box across a horizontal floor, or when a horse-drawn barge is pulled along a narrow canal using a rope from the bank of the canal, the motion is restricted by contact forces.

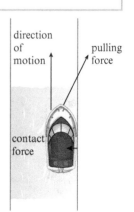

When the direction of a force is different from the direction of motion, we can calculate the work done by the force by resolving it into components along the direction of motion and perpendicular to the direction of motion.

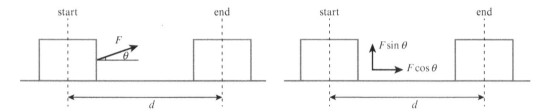

The box in the diagram is on the floor and there is no motion in the perpendicular direction, so the perpendicular component does no work. The work done by the force is given by the component of the force in the direction of motion multiplied by the distance moved.

KEY POINT 8.2

When a force of magnitude F N moves a body a distance d m, at an angle θ to the direction of the force, the work done by the force is:
$$W = Fd \cos \theta$$

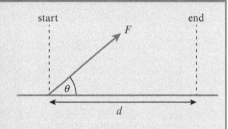

REWIND

Look back to Chapter 3, Section 3.1, if you need a reminder about resolving forces into perpendicular components.

You can either think of this as the component of the force in the direction of motion multiplied by the distance moved, $W = (F \cos \theta) \times d$, as in the following left-hand diagram; or as the force multiplied by the component of the displacement in the direction of the force, $W = F \times (d \cos \theta)$, as in the following right-hand diagram.

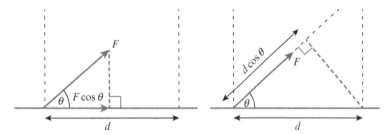

WORKED EXAMPLE 8.4

A small truck is pulled 5 m along a railway track by a force of 100 N at an angle 60° to the track. Find the work done by the force.

Answer

Work done $= Fd \cos \theta = 100 \times 5 \times \cos 60$
$= 250$ J

We can think of this as 50 N × 5 m or as 100 N × 2.5 m.

If there is a force that opposes the direction of motion, then the work done by this force will be negative and we say that work is done *against* the force. This happens, for example, when a load is being raised vertically and work is being done against its weight (or against gravity), as in the next Worked example.

WORKED EXAMPLE 8.5

A ball of mass 0.05 kg is raised through a distance of 1.5 m. Find the work done against gravity.

Answer

Weight = $0.05 \times 10 = 0.5$ N

> There will be other forces acting to raise the ball, but we are only asked about the work done against gravity; that is, due to the weight.

Work done by the weight = $0.5 \times -1.5 = -0.75$ J

> This is negative because the weight opposes the motion.

Work done against the weight = −(work done by the weight)
Work done against weight = 0.75 J
Work done against gravity = 0.75 J

> We can say that we have negative work done *by* the weight or that we have positive work done *against* the weight.

When several forces act on a body we can add the work done by each force to get the total work done by all the forces. Remember that work done may be positive or negative. So to find the total work done by the forces, we add the work done by the forces with components in the direction of motion (forces that help to move the body) and subtract the work done against forces with components in the direction opposite to the direction of motion (forces that try to prevent the body from moving). This is illustrated in Worked example 8.6.

WORKED EXAMPLE 8.6

A box of mass 5 kg is pushed up a slope inclined at 30° to the horizontal by a force of 30 N at an angle 10° to the slope. The frictional force acting on the box is 2 N. The box moves a distance 3 m up the slope.

  a Find the work done against friction.
  b Find the work done against gravity.
  c Find the work done by the push force.
  d Find the work done by the normal reaction.
  e Find the total work done on the box by all four forces.

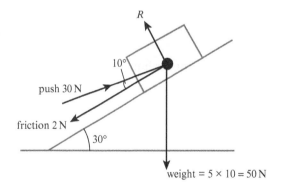

Answer

  a Work done against friction = $2 \times 3 = 6$ J

> Friction acts along the direction of motion, but opposing the motion.

Cambridge International AS & A Level Mathematics: Mechanics

b The component of the weight down the slope is

50 sin 30 = 25 N

Work done against gravity = 25 × 3 = 75 J

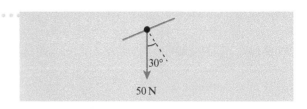

The angle between the push force and the slope is 10°.

c The component of the push force up the slope is

30 cos 10 = 29.54 N

Work done by push force = 29.54 × 3 = 88.6 J

d There is no movement in the perpendicular direction, so no work is done by the normal reaction.

e Total work done = work done by push force
− work done against gravity
− work done against friction
= 88.6 − 75 − 6 = 7.6 J

This is the total work done by all four forces in moving the box.

TIP

Sometimes a question may mention 'non-gravitational resistance'. This means all the components of forces, such as friction and air resistance, that act against the motion. It does not mean any component of the weight that would oppose the motion of a body travelling uphill or rising vertically.

EXERCISE 8A

1 A crate is pushed 2 m across a smooth horizontal floor by a horizontal force of 30 N. Find the work done by the force.

2 A box is pulled 5 m across a smooth horizontal floor by a rope with tension 20 N. Find the work done by the tension:

   a when the rope is horizontal

   b when the rope is at 40° above the horizontal.

3 A ball of mass 0.04 kg is thrown vertically upwards. It rises 2 m and then falls 2 m. Ignoring air resistance, find the work done by gravity:

   a when the ball rises 2 m

   b when the ball falls 2 m

   c when the ball rises 2 m and then falls 2 m.

4 A skier of mass 60 kg starts from rest at the top of a slope of vertical height 10 m. She descends the slope and ascends the other side to come to rest at a point that is 4 m vertically lower than where she started. Find the total work done by gravity (i.e. the work done by gravity in descending minus the work done against gravity while ascending).

5 A horse-drawn barge is pulled 20 m forwards using a rope at a small angle to the direction of motion. The barge touches against the edge of the canal. The total resistance to the motion is 100 N.

   a What causes the resistance?

   b Determine the work done against the resistance.

6 A horse-drawn barge is pulled 40 m forwards using a rope at an angle θ to the direction of motion. The tension in the rope is 150 N. The barge is kept moving in a straight line by a contact force with the edge of the canal. Resistance forces also act on the barge.

   a Determine the work done by the tension:

      i   when $\theta = 10°$

      ii when $\theta = 20°$.

   b Show that when $\theta = 20°$ the tension would need to increase to 157.2 N to do the same work as in part **ai**.

   Consider the barge being pulled with a tension of 150 N with $\theta = 10°$, or being pulled with a tension of 157.2 N with $\theta = 20°$.

   c Explain why the frictional resistance will be greater in the second of these situations.

7 A box is pulled 2 m across a horizontal floor, using a rope with tension 10 N at 30° to the horizontal. The frictional resistance is 3 N. Find:

   a the work done against friction

   b the work done by the tension

   c the work done by the weight

   d the work done by the normal contact force

   e the total work done by all four forces.

8 A crate of mass 25 kg slides 4 m down a slope that is inclined at 15° to the horizontal. Non-gravitational resistance is 5 N. Find:

   a the work done against non-gravitational resistance

   b the work done against gravity

   c the work done by the normal contact force

   d the total work done by all these forces.

9 A crate of mass 25 kg is pushed 4 m up a slope that is inclined at 15° to the horizontal by a force of 100 N parallel to the slope. Non-gravitational resistance is 5 N. Find:

   a the work done by the force of 100 N

   b the work done against non-gravitational resistance

   c the work done against gravity

   d the work done against the normal contact force

   e the total work done by all these forces.

10 A tile of mass 0.5 kg slides 2 m down a roof, which is inclined at 60° to the vertical. The frictional force is 1.5 N. There are no other external forces. Find:

   a the work done by gravity

   b the work done against friction

   c the work done by the normal reaction force

   d the total work done by all three forces.

PS 11 A box of weight 20 N is pulled 12 m across a horizontal floor, using a rope with tension 25 N at 10° to the horizontal. The frictional resistance is 1 N. Find the total work done by all three forces.

PS 12 A sack of mass 5 kg slides 2 m down a ramp. The ramp is inclined at 15° to the horizontal. The coefficient of friction between the sack and the ramp is 0.25. Find the total work done.

8.2 Kinetic energy

Energy can exist in many forms: heat, light, nuclear energy, chemical energy (from food or fuel), stored (potential) energy (such as the energy stored in a compressed spring) and so on. Energy can be transferred from one form to another and can be used to create motion.

Energy is a scalar quantity; it can be positive or negative but otherwise has no direction.

In Mechanics we are only interested in mechanical energy. Mechanical energy can be kinetic or potential.

Kinetic energy is the energy that a body possesses because of its motion.

KEY POINT 8.3

A body of mass m kg moving with speed v m s^{-1} has kinetic energy (KE) given by:

$$KE = \frac{1}{2} mv^2$$

Kinetic energy could be measured in $(\text{kg})(\text{m s}^{-1})^2$, but this is the same as $(\text{kg m s}^{-2})(\text{m}) = \text{N m} = \text{J}$.

All forms of energy are measured in joules (J).

WORKED EXAMPLE 8.7

Find the kinetic energy of a body of mass 3 kg moving at 5 m s^{-1}.

Answer

$KE = \frac{1}{2} mv^2 = \frac{1}{2} \times 3 \times 5^2$ Substitute the values for m and v into the formula for the kinetic energy.

$= 37.5 \, \text{J}$ Remember to give units.

Chapter 8: Work and energy

WORKED EXAMPLE 8.8

A ball of mass 50 g hits the ground with speed $10\,\text{m s}^{-1}$ and rebounds with speed $6\,\text{m s}^{-1}$. Find the loss in kinetic energy that occurs in the bounce.

Answer

$50\,\text{g} = 0.050\,\text{kg}$ .... Convert the mass to kg.

Change in KE = final KE − initial KE

KE before $= \dfrac{1}{2} \times 0.050 \times 10^2 = 2.5\,\text{J}$ .... Calculate the KE just before the bounce. Initial KE $= \dfrac{1}{2}mu^2$

KE after $= \dfrac{1}{2} \times 0.050 \times 6^2 = 0.9\,\text{J}$ .... Calculate the KE just after the bounce. Final KE $= \dfrac{1}{2}mv^2$

KE is scalar, so the change in the direction of the velocity does not matter.

So loss in KE $= 1.6\,\text{J}$ .... Remember to give units.

KEY POINT 8.4

A common error is to use the difference between the velocities or speeds in the calculation, like this:

$$\dfrac{1}{2} \times 0.050 \times (6+10)^2 = 6.4\,\text{J} \quad \text{or} \quad \dfrac{1}{2} \times 0.050 \times (6-10)^2 = 0.4\,\text{J}$$

But these are both wrong. It must be the difference of the squares of the speeds:

$$\dfrac{1}{2} \times 0.050 \times (6^2 - 10^2) = 1.6\,\text{J}$$

EXERCISE 8B

1. Find the kinetic energy of an object of mass 10 kg moving at $8\,\text{m s}^{-1}$.
2. Find the kinetic energy of a car of mass 1500 kg moving at $22\,\text{m s}^{-1}$.
3. Find the kinetic energy of a tennis ball of mass 57 g moving at $180\,\text{km h}^{-1}$.
4. A rock of mass 4 kg is thrown upwards with an initial speed of $3\,\text{m s}^{-1}$. It is travelling at $6\,\text{m s}^{-1}$ just before it lands. Find the increase in its kinetic energy.
5. A book of mass 2 kg falls from a window ledge and drops 10.8 m to the ground. It falls freely under gravity.
   a Find the speed of the book just before it hits the ground.
   b Find the kinetic energy of the book just before it hits the ground.
6. A model train of mass 1 kg is moving at $3\,\text{m s}^{-1}$. It accelerates uniformly for 5 s, travelling in a straight line and covering a distance of 40 m while accelerating.
   a Find the speed of the train at the end of the 5 s.
   b Find the increase in kinetic energy from the start to the end of the 5 s.

7 A ball bearing with mass 0.03 kg is projected vertically upwards. It loses 0.735 J of kinetic energy before coming to instantaneous rest. Find the initial speed of the ball bearing.

8 A box of mass 30 kg slides from the top of a smooth slope to the bottom of the slope. The slope is inclined at 30° to the horizontal. The box starts from rest. At the bottom of the slope the box has gained 375 J of kinetic energy. Find the length of the slope.

9 A boy of mass 64 kg runs at a constant speed along a straight track. He takes 16 s to run 100 m.
   a Work out his kinetic energy.
   b What difference would it make if the track was curved?

10 At its launch a rocket has mass 2 million kg. It accelerates from rest to $75\,000\,\text{m s}^{-1}$.
   a Work out the increase in the kinetic energy.
   b Why will the calculated value be too big?

11 Ball A, of mass 2 kg, is moving in a straight line at $5\,\text{m s}^{-1}$. Ball B, of mass 4 kg, is moving in the same straight line at $2\,\text{m s}^{-1}$. Ball B is travelling directly towards ball A. The balls hit each other and after the impact each ball has reversed its direction of travel. The kinetic energy lost in the impact is 12.5 J.
   a Show that the speed of ball A after the impact is $\dfrac{10}{3}\,\text{m s}^{-1}$.
   b Find the speed of ball B after the impact.

12 Two balls, A and B, of equal mass, are travelling towards one another with velocities u_A and $-u_B$, respectively. The balls collide and their velocities after the impact are $-v_A$ and v_B, respectively. The kinetic energy after the impact is the same as the kinetic energy before the impact (i.e. a 'perfectly elastic collision'). Explain why $v_A = u_B$ and $v_B = u_A$.

13 Balls X, Y and Z lie at rest on a smooth horizontal surface, with Y between X and Z. Balls X and Z each have mass 2 kg and ball Y has mass 1 kg. Ball X is given a velocity of $1\,\text{m s}^{-1}$ towards ball Y. Balls X and Y collide. After this collision the speed of ball X is $0.4\,\text{m s}^{-1}$ in its original direction.
   a Work out the loss in kinetic energy in this impact.

   Ball Y goes on to collide with ball Z. After this collision the speed of ball Y is $0.4\,\text{m s}^{-1}$ in the direction towards ball X.
   b Work out the loss in kinetic energy in this impact.

   Finally, ball Y collides with ball X again. After this collision the speed of ball Y is twice the speed of ball X and the speed of ball Z is four times the speed of ball Y, with the balls all travelling in the same direction.
   c Work out the loss in kinetic energy in this impact.

EXPLORE 8.1

Investigate what happens in Exercise 8B, question 12, if the perfectly elastic collision takes place between two balls that have different masses.

8.3 Gravitational potential energy

The other type of mechanical energy is potential energy. Potential energy is the energy that a body possesses because of its position. It can be thought of as stored energy.

Gravitational potential energy is the energy that could be released if the body falls under gravity.

> **TIP**
>
> Gravitational potential energy (GPE) is sometimes just called 'potential energy', although there are other types of potential energy (e.g. elastic potential energy, which is the energy stored in a stretched or compressed spring).

KEY POINT 8.5

A body of mass m kg at height h m has potential energy (PE) given by:

$$PE = mgh = 10mh$$

The amount of potential energy that a body possesses depends on its height. The height is measured from some 'base' level where PE = 0. We can choose any level as the base, but must measure all heights from the same level.

Potential energy is measured in joules (J).

WORKED EXAMPLE 8.9

Find the increase in potential energy in raising a sack of mass 1 kg through a height of 3 m.

Answer

Increase in PE $= mgh = 1 \times 10 \times 3$ ⋯ Substitute the values for m, g and h into the formula for PE.

$= 30$ J ⋯ Remember to give units.

When a body of mass m kg is raised through a height h m, the work done against gravity is mgh J and the increase in gravitational potential energy is mgh J. Potential energy increases when work is done against gravity, so objects at a higher level have more potential energy than those that are lower.

When the same body descends through a vertical distance h m, the work done by gravity is mgh J and the decrease in gravitational potential energy is mgh J. Potential energy decreases when work is done by gravity.

What matters is the *vertical height* difference between the top and the bottom, even if the body descends by sliding down a slope.

MODELLING ASSUMPTIONS

As always, we are assuming objects are particles. This means that when we consider the kinetic energy of an object, we assume the entire object is moving at the same speed. This is often not the case. For example, the wheels of a car are rotating, so the point at the top of the wheel is moving more quickly than the car, but the point at the bottom is moving more slowly. We will consider this difference as negligible. Experimenting with a ball rolling down a slope will show that the speed at the bottom of the slope is not as high as expected.

EXERCISE 8C

1. Find the increase in the potential energy of an object of mass 5 kg when it rises through 2 m.

2. Find the change in the potential energy of a body of mass 10 kg when it falls through a height of 6 m, stating whether this is an increase or a decrease.

3. Find the increase in the potential energy of a tennis ball of mass 57 g when it rises through a height of 70 cm.

4. A box of mass 25 kg falls 2 m vertically downwards. Find:
   a the loss in potential energy
   b the work done by gravity
   c the increase in kinetic energy.

5. A tile of mass 1.2 kg slides 3 m down a roof that makes an angle of 35° to the horizontal. Find the decrease in potential energy.

6. A person of mass 70 kg climbs three flights of stairs to reach the third floor of a building. Each of the flights of stairs consists of 15 stairs, each of depth 18 cm. Find the increase in the potential energy of the person when they climb from the ground floor to the third floor.

7. A crate is pulled up a smooth slope using a rope that is parallel to the slope. The slope is inclined at an angle θ to the horizontal, where $\sin\theta = 0.28$, and the tension in the rope is 50 N. The work done by the tension is 75 J and the increase in the potential energy of the crate is 168 J. Find the mass of the crate.

8. A ramp rises 10 cm for every 80 cm along the sloping surface. A box of mass 50 kg slides down the ramp, starting from rest at the top of the ramp. The coefficient of friction between the ramp and the box is 0.03 and no other resistance forces act.
   a Draw a diagram to show the forces acting on the box.

   The box is travelling at $2\,\text{m s}^{-1}$ when it reaches the bottom of the ramp.

   b Find the length of the ramp.
   c Find the loss in the potential energy of the box.

9. A boy of mass 60 kg slides down a slope that makes an angle of 45° to the horizontal. The coefficient of friction between the boy and the surface of the slope is 0.2. The boy starts from rest at the top of the slope and finishes at the end with speed $v\,\text{m s}^{-1}$.
   a Show that the acceleration of the boy is $5.66\,\text{m s}^{-2}$ down the slope.
   b Work out the length of the slope in terms of v.
   c Find an expression for the loss in the boy's potential energy in terms of v when he slides down the slope.
   d What modelling assumptions have you made and what effect would each of these have on your answer to part c?

PS 10 The recommended slope for wheelchair ramps is 1:12. This means that the ramp rises 1 cm vertically for every 12 cm along the slope.

A person in their wheelchair, with total mass 90 kg, descend along a 1:12 wheelchair ramp that has slope length 8 m.

They start the descent with speed 2 m s^{-1} and finish with speed 4 m s^{-1}.

Work out the change in total mechanical energy (kinetic energy + potential energy) after descending the ramp.

PS 11 A ball of mass 0.02 kg is projected vertically upwards through oil. The ball has initial speed $\frac{35}{12} \text{ m s}^{-1}$. The oil exerts a resistance of $0.1t^{0.5}$ N, where t s is the time from when the ball was projected.

Work out the increase in the potential energy of the ball from the start to when it comes to instantaneous rest. (Note: you will need an equation solver for this question. You will not be allowed an equation solver in the examination.)

P 12 A particle of mass m kg is projected up a slope. The particle has initial speed $v \text{ m s}^{-1}$ up the slope. The slope is inclined at an angle θ to the horizontal and the coefficient of friction between the particle and the slope is μ.

Show that when the particle comes to rest its potential energy has increased by $\dfrac{mv^2 \tan \theta}{2(\mu + \tan \theta)}$.

Checklist of learning and understanding

- The work done, in joules, by a force of magnitude F N in moving a body a distance d m in the direction of the force is:

  $W = Fd$

- The work done, in joules, by a force of magnitude F N in moving a body a distance d m at an angle θ to the direction of the force is:

  $W = Fd \cos \theta$

- The kinetic energy, in joules, of a body of mass m kg moving with speed $v \text{ m s}^{-1}$ is:

  $\text{KE} = \frac{1}{2} mv^2$

- The gravitational potential energy, in joules, of a body of mass m kg at height h m above the base level is:

  $\text{GPE} = mgh$

END-OF-CHAPTER REVIEW EXERCISE 8

1. A block is pulled for a distance of 50 m along a horizontal floor, by a rope that is inclined at an angle of $\alpha°$ to the floor. The tension in the rope is 180 N and the work done by the tension is 8200 J. Find the value of α. [3]

   Cambridge International AS & A Level Mathematics 9709 Paper 43 Q1 June 2011

2. A ball of mass 30 g is thrown vertically upwards with an initial speed of $4\,\text{m s}^{-1}$. Air resistance can be ignored. The ball reaches a maximum height of 80 cm. Find:

   a the decrease in kinetic energy [2]

   b the increase in potential energy. [2]

3. A car of mass 1600 kg is driven 200 m along a straight horizontal road. The car starts with a speed of $3\,\text{m s}^{-1}$ and finishes with a speed of $20\,\text{m s}^{-1}$. A constant resistance of 40 N acts.

   a Find the acceleration of the car. [2]

   b Find the work done by the driving force. [2]

4. A box of mass 20 kg is pushed 3 m up a slope inclined at an angle α to the horizontal. The work done by the push force is 900 J and non-gravitational resistance (friction and air resistance) is 40 N.

   a Work out the push force. [1]

   b Show that the acceleration of the box up the slope, $a\,\text{m s}^{-2}$, is given by $a = 13 - 10 \sin \alpha$. [3]

   c What assumptions have you made? [1]

5. A and B are two points 50 metres apart on a straight path inclined at an angle θ to the horizontal, where $\sin \theta = 0.05$, with A above the level of B. A block of mass 16 kg is pulled down the path from A to B. The block starts from rest at A and reaches B with a speed of $10\,\text{m s}^{-1}$. The work done by the pulling force acting on the block is 1150 J.

   i Find the work done against the resistance to motion. [3]

   The block is now pulled up the path from B to A. The work done by the pulling force and the work done against the resistance to motion are the same as in the case of the downward motion.

   ii Show that the speed of the block when it reaches A is the same as its speed when it started at B. [2]

   Cambridge International AS & A Level Mathematics 9709 Paper 42 Q2 June 2013

6. A basketball of mass 0.625 kg is thrown from a height of 2 m with a speed of $6\,\text{m s}^{-1}$. It passes through the hoop at a height of 3 m with speed $4\,\text{m s}^{-1}$.

   a Find the change in the kinetic energy of the ball, stating whether this is an increase or a decrease. [3]

   b Find the change in the potential energy of the ball, stating whether this is an increase or a decrease. [3]

   c What difference does changing the ball's angle of projection make? [1]

7. A lorry of mass 16 000 kg moves on a straight hill inclined at angle $\alpha°$ to the horizontal. The length of the hill is 500 m.

   i While the lorry moves from the bottom to the top of the hill at constant speed, the resisting force acting on the lorry is 800 N and the work done by the driving force is 2800 kJ. Find the value of α. [4]

   ii On the return journey the speed of the lorry is $20\,\text{m s}^{-1}$ at the top of the hill. While the lorry travels down the hill, the work done by the driving force is 2400 kJ and the work done against the resistance to motion is 800 kJ. Find the speed of the lorry at the bottom of the hill. [4]

   Cambridge International AS & A Level Mathematics 9709 Paper 43 Q5 June 2012

8 A box of mass 20 kg moves across a horizontal floor. The coefficient of friction between the box and the floor is 0.2, and friction is the only resistance force. The box has initial speed $3\,\text{m s}^{-1}$ and moves until it comes to rest.
   a Find the retardation (negative acceleration) of the box. [3]
   b Find the distance that the box travels. [3]
   c Find the work done against friction. [3]

9 Sam and his skateboard have a combined mass of 100 kg. He accelerates from $0\,\text{m s}^{-1}$ to $20\,\text{m s}^{-1}$ while descending a hill. The hill is modelled as a slope at an angle of $\sin^{-1} 0.2$ to the horizontal. The non-gravitational resistance is 200 N. The bottom of the hill is 10 m below the top of the hill. Find:
   a the increase in the kinetic energy of Sam and his skateboard [3]
   b the decrease in the potential energy of Sam and his skateboard [2]
   c the distance that the skateboard travels [3]
   d the work done against resistance. [2]

10 Kiera climbs up a ladder to sit at the top of a slide 2 m above the ground. Her potential energy increases by 1280 J.
   a Find Kiera's weight. [1]

   Kiera then slides down the slide, starting from rest. The slide is modelled as a slope at an angle θ to the horizontal. The resistance force is a constant 20 N. The work done against resistance by Kiera when she is sliding is 80 J.
   b Find the length of the slide. [2]
   c Find the value of θ. [3]
   d Find Kiera's speed when she reaches the bottom of the slide. [4]

11 A ramp is inclined at an angle $\sin^{-1} 0.1$ to the horizontal. A box of mass 40 kg is projected up the ramp with initial speed $5\,\text{m s}^{-1}$. The coefficient of friction between the ramp and the box is 0.05, and no other resistance forces act.
   a Find the acceleration of the box, stating its direction. [4]

   The box comes to rest when it reaches the top of the ramp.
   b Find the length of the ramp. [3]
   c Find the gain in the potential energy of the box. [3]

   The total mechanical energy is the sum of the kinetic energy and the potential energy.
   d Show that the overall loss in the mechanical energy of the box is 166 J. [3]

12 Jack has mass 70 kg. He works as a 'human cannon ball'. Jack is projected with speed $12\,\text{m s}^{-1}$ at an angle of $45°$ above the horizontal. He lands on a trampoline when the angle between his flight and the horizontal is $50°$. Model Jack as a particle with no air resistance.
   a Explain why the horizontal component of Jack's velocity is constant. [2]
   b Find Jack's speed when he hits the trampoline. [4]
   c Find the kinetic energy gained during the flight. [3]

   The gain in Jack's kinetic energy equals the loss in his gravitational potential energy.
   d Find the difference in height between the mouth of the cannon and the trampoline. [3]

   By changing the angle of projection, Jack can change the angle between his flight and the horizontal when he lands. Suppose that Jack lands on the trampoline at an angle α to the horizontal.
   e What could happen if α is very small? [1]
   f What could happen if α is close to $90°$? [1]

Chapter 9
The work–energy principle and power

In this chapter you will learn how to:

- use the work–energy principle
- understand when mechanical energy is conserved
- calculate the power of a moving body
- use power to calculate the maximum speed of a moving body.

Chapter 9: The work–energy principle and power

PREREQUISITE KNOWLEDGE

Where it comes from	What you should be able to do	Check your skills
Chapter 8	Calculate kinetic energy.	1 A box of mass 5 kg is pushed up a slope. The box has initial speed $2\,\text{m s}^{-1}$ and final speed $3\,\text{m s}^{-1}$. Find the increase in the kinetic energy of the box.
Chapter 8	Calculate the work done by a force in moving a body.	2 A box of mass 5 kg is pushed 5 m up a slope inclined at 30° to the horizontal by a force of 30 N parallel to the slope. The frictional force acting on the box is 3 N. a Find the work done by the push force. b Find the work done against friction. c Find the work done against gravity.

How is power used in mechanics?

We talk about a 'powerful argument' to mean a persuasive argument, or a 'power lifter' as someone who lifts great weights. Political activists talk about giving 'power to the people' when they mean giving rights to a group of people or acting on the wishes of the majority. In everyday use the word power means something like strength, but in mechanics the word strength relates to the force needed to break something (such as the breaking strength of a cable). Power in mechanics is a way of measuring the rate at which a machine generates motion.

In this chapter you will learn how energy can be converted from one form to another and how work can increase or decrease the mechanical energy (kinetic and gravitational potential energy) of a body. You will also learn how the relationship between the power generated by the engine of a vehicle and the work that is done by the driving force lets us find the maximum speed that can be achieved by the vehicle.

9.1 The work–energy principle

When a force moves a body, it does work and causes a change in the kinetic energy of the body.

For motion in a straight line with constant acceleration we know that

$$v^2 = u^2 + 2as$$

Using Newton's second law we can replace a by $\dfrac{F}{m}$ to give

$$v^2 = u^2 + 2\dfrac{F}{m}s$$

Multiplying by $\dfrac{1}{2}m$ and rearranging gives

$$\dfrac{1}{2}mv^2 - \dfrac{1}{2}mu^2 = Fs$$

You know from Chapter 8 that $\dfrac{1}{2}mv^2 - \dfrac{1}{2}mu^2$ is the increase in kinetic energy. When the motion is in a straight line and in a constant direction, the distance moved will be the same as the displacement, s, and so the work done by the force is given by $W = Fs$.

TIP

If $|v| < |u|$, then $\dfrac{1}{2}mv^2 - \dfrac{1}{2}mu^2$ is negative and there is a decrease in kinetic energy.

Cambridge International AS & A Level Mathematics: Mechanics

This means that the previous equation can be expressed as:

increase in kinetic energy = work done by force

This relationship between work done and kinetic energy is not restricted to motion in a straight line, nor to motion with constant acceleration. We do not need to know the exact path taken to get from the start to the finish. This means that we can easily deal with non-linear motion or situations where we know what happens at the start and at the finish but not the exact path taken in between.

This result tells us how the forces acting cause the kinetic energy to increase or decrease. The total work done by all the forces acting (driving force, weight, non-gravitational resistance etc.) equals the increase in kinetic energy.

> **TIP**
>
> The path of the body can be any curve, or even unknown. For example, the work–energy principle applies to a child on a slide, helter-skelter or roller coaster; a person skiing in a zigzag path or up and down hills; or the motion of a particle moving in a circle.
>
> Application of the work–energy principle is the only method that can be used when the path is not a straight line.

KEY POINT 9.1

The **work–energy principle** states that for *any* motion:

increase in kinetic energy = total work done by all forces

$$\frac{1}{2}mv^2 - \frac{1}{2}mu^2 = \sum Fs$$

where the 'total work done' is the sum of the work done by forces (including weight) with a component in the direction of motion (forces that speed up the motion) minus the work done against forces with a component in the direction opposing the motion (forces that slow down the motion).

The total work done will include work done by any force that is not perpendicular to the direction of motion. This includes any driving force, push or pull, tension or compression, weight, air resistance, friction etc.

The work–energy principle applies whatever the path taken during the motion.

WORKED EXAMPLE 9.1

A boy uses a constant force of 250 N to push a box, of mass 20 kg, a distance 4 m in a curved path across a horizontal floor. The box starts from rest. Find the final speed of the box:

  a when the floor is smooth

  b when the coefficient of friction between the floor and the box is 0.12.

Answer

a Work done $= 250 \times 4$

                    $= 1000$ J   *The only force that does work is the push force.*

  Using the work–energy principle: *The path is not a straight line, so you need to use the work–energy principle.*

  $\frac{1}{2}mv^2 - \frac{1}{2}mu^2 = $ work done

  $u = 0$ *The box starts from rest.*

  $\frac{1}{2}mv^2 - 0 = 1000$

$v^2 = 1000 \div 10$
$= 100$
$v = 10$

So the final speed of the box is $10\,\text{m s}^{-1}$.

$\frac{1}{2}m = 10$

b Work done by push force = 1000 J

Work is done by the push force and work is done against friction.

Friction force = 0.12×200
$= 24\,\text{N}$

so work done against friction = 24×4
$= 96\,\text{J}$

Total work done = WD by push force − WD against friction
$= 1000 - 96 = 904\,\text{J}$

'WD' = work done

$u = 0$

$\frac{1}{2}mv^2 - 0 = 904$

$v^2 = 904 \div 10$
$= 90.4$
$v = 9.51$

So final speed is $9.51\,\text{m s}^{-1}$.

When the motion involves a change in the height of the body, work will be done by or against the weight of the body.

The total work done can be written as the sum of the work done by the weight and the work done by the other forces.

When the height of a body increases, the work done against the weight (or against gravity) is the same as the increase in gravitational potential energy; when the height decreases, the work done by the weight (or by gravity) is the same as the decrease in gravitational potential energy.

We have:

increase in kinetic energy = total work done

and:

total work done = work done by the weight + total work done by the other forces
= decrease in gravitational potential energy + total work done by other forces

This gives an alternative form for the work–energy principle:

increase in kinetic energy + increase in gravitational potential energy = total work done by forces

(where 'forces' here excludes the weight of the body).

The sum of the kinetic energy and the gravitational potential energy is the total mechanical energy.

TIP

Kinetic and potential energy are types of mechanical energy. Other forms of energy (heat, light, sound, chemical, electrical, nuclear etc.) are non-mechanical.

Cambridge International AS & A Level Mathematics: Mechanics

> **KEY POINT 9.2**
>
> We can write the work–energy principle as
>
> increase in mechanical energy = total work done by forces that act to speed the body up
> − total work done by forces that act to slow the body down
>
> (in both cases 'forces' excludes the weight of the body).

WORKED EXAMPLE 9.2

A ball of mass 0.05 kg is thrown vertically upwards with an initial speed of u m s^{-1}. It rises through a distance of 1.5 m and then falls through 2.5 m before hitting the floor. It hits the floor with speed v m s^{-1}. Throughout the motion air resistance of 0.01 N acts on the ball. Calculate the initial speed, u m s^{-1}, and the final speed, v m s^{-1}.

Answer

Using the work–energy principle: | State that you are using the work–energy principle.

increase in mechanical energy = work done

increase in KE + increase in GPE = 0 − WD against resistance

To find u we consider the motion from the start to the top:

increase in KE $= 0 - \frac{1}{2} \times 0.05 \times u^2$

$= -0.025u^2$ J | The KE decreases by $0.025u^2$ J.

increase in GPE $= 0.05 \times 10 \times 1.5$

$= 0.75$ J

so increase in mechanical energy $= 0.75 - 0.025u^2$ J

Work done against resistance $= 0.01 \times 1.5 = 0.015$ J

$0.75 - 0.025u^2 = -0.015$ | Increase in mechanical energy = −WD against resistance, so it is negative (it is a decrease).

$u^2 = 30.6$

$u = 5.53$

Initial speed $= 5.53$ m s^{-1}

To find v we consider the motion from the top to the floor: | Use the work–energy principle again for the second part of the motion.

increase in KE $= \frac{1}{2} \times 0.05 \times v^2 - 0 = 0.025v^2$ J

increase in GPE $= 0.05 \times 10 \times -2.5$

$= -1.25$ J

So increase in mechanical energy from the top to the floor

$= (0.025v^2 - 1.25)$ J

Work done against resistance $= 0.01 \times 2.5$ | Air resistance is constant throughout the motion.

$= 0.025$ J

$0.025v^2 - 1.25 = -0.025$

$0.025v^2 = 1.225$

$v = 7$

Final speed = 7 m s^{-1}

> Increase in mechanical energy = −WD against resistance.

Alternatively, we could consider the entire motion together.

From the start to the end:

increase in KE = $0.025(v^2 - u^2)$
$= 0.025v^2 - 0.765$

increase in GPE = $0.05 \times 10 \times (1.5 - 2.5)$
$= -0.5 \text{ J}$

So increase in mechanical energy = $(0.025v^2 - 1.265) \text{ J}$

Work done against resistance = $0.01 \times (1.5 + 2.5)$
$= 0.04 \text{ J}$

> Note that the resistance acts for a total distance of 4 m of travel, although the displacement is only 1 m downwards.

$0.025v^2 - 1.265 = -0.04$

$0.025v^2 = 1.225$

$v = 7 \text{ m s}^{-1}$

WORKED EXAMPLE 9.3

A woman snowboards down a hill of varying gradient. The mass of the woman and her snowboard is 64 kg. She starts from rest at the top of the hill and accelerates under gravity. Throughout the descent the woman does no work to accelerate or decelerate the snowboard, the average frictional force is 1.5 N and all other resistance forces are negligible. The snowboarder reaches the bottom of the hill with a speed of 30 m s^{-1}, having travelled a distance of 500 m in a zigzag route down the hill. Find the height of the hill, h metres.

Answer

Using the work–energy principle:

increase in mechanical energy = work done

increase in KE + increase in PE = 0 − WD against friction

> The motion is non-linear so the work–energy principle must be used.

Initial speed = 0 m s^{-1}

> The snowboarder starts from rest.

Final speed = 30 m s^{-1}

So increase in kinetic energy = $\frac{1}{2} \times 64 \times 30^2$
$= 28\,800 \text{ J}$

Initial gravitational potential energy = 64 × 10 × h .... Take the bottom of the hill as the zero level
Final gravitational potential energy = 0   for potential energy.
So increase in potential energy = −640h J

The work done against friction = 1.5 × 500 .... (Average) force × distance
                               = 750 J

Hence, 28 800 − 640h = −750 .... Substitute the values into the work–energy
       h = 46.2   equation.

The height of the hill is 46.2 m.

EXERCISE 9A

1 A box of mass 25 kg is pulled 5 m across a smooth floor by a rope with tension 22 N. The rope is horizontal. There is a frictional force with average value 12 N. The box starts from rest. Find:

   a the work done against friction

   b the work done by the tension

   c the total work done by all the forces

   d the final speed of the box.

2 For the situation described in question 1, find the final speed when the rope is inclined at 40° above the horizontal.

3 A crate of mass 50 kg slides down a smooth slope. At the top of the slope the crate has speed $0\,\mathrm{m\,s^{-1}}$ and at the bottom of the slope it has speed $4\,\mathrm{m\,s^{-1}}$.

   Find:

   a the increase in kinetic energy

   b the decrease in potential energy

   c the vertical height through which the crate has descended.

4 A boy sledges down a hill. The boy and his sledge have a combined mass of 85 kg. He starts from rest and descends through a vertical height of 3 m. Friction and air resistance are negligible.

   a Find the work done by gravity.

   b Use the work–energy principle to find the boy's speed at the end of the descent.

   The boy descends the hill again, starting from rest, but this time he is joined on the sledge by his little brother, of mass 35 kg.

   c Find their speed at the end of the descent.

5 A girl of mass 50 kg travels down a water slide. She starts at the top with a speed of $2\,\mathrm{m\,s^{-1}}$ and descends through a vertical height of 5 m.

   a Assuming that there is no resistance, find her speed when she reaches the bottom of the slide.

   b The girl's actual final speed is $8\,\mathrm{m\,s^{-1}}$ because there is resistance of average value 40 N. Find the length of the water slide.

6 A child of mass 45 kg travels down a water chute. The child has speed 1 m s⁻¹ at the top of the chute and speed 5 m s⁻¹ at the bottom of the chute. The length of the water chute is 20 m and the height through which it descends is 4 m. Work out the average resistance force that acts.

7 A boy sits on a sledge at the top of an icy hill. He gently sets the sledge in motion. When he reaches the bottom of the hill he is moving at 10 m s⁻¹. Assuming that friction is negligible, find the height of the hill.

8 A girl of mass 50 kg sits on a sledge at the top of a grassy hill. She gently sets the sledge in motion. When she reaches the bottom of the hill she is moving at 9.9 m s⁻¹. The hill is 5 m high and the sledge slides 100 m down the hill.

   a Work out the resistance force.
   b Comment on your answer.

9 A child of mass 40 kg slides down a playground slide. The child starts from rest at the top of the slide, 2 m above the ground. At the bottom of the slide its slope levels off.

   a Find the child's loss of gravitational potential energy.

   There is a constant resistance of 112 N throughout.

   b Find the distance the child has travelled when she comes to rest.

   The slide is inclined at an angle of 30° to the horizontal.

   c Find the distance the child travels on the level part of the slide.

10 A car of mass 1600 kg travels 200 m along a level road. The average driving force is 2000 N and the average resistance is 800 N. The driver claims that the speed throughout the journey was less than 30 m s⁻¹. What can you say about the initial speed of the car?

11 A roller-coaster car has mass 100 kg. It carries two passengers, each of mass between 50 kg and 80 kg. The car becomes detached from the drive chain and continues to travel along the ride with no drive force and no braking force. The car comes to instantaneous rest at the highest point of the ride and then descends under gravity to reach the lowest point of the ride. The highest point is 12 m vertically above the lowest point. The car travels 100 m along the track while descending through 12 m. When the car passes through the lowest point it has speed 15 m s⁻¹.

   a Show that the average frictional force is less than 20 N.
   b If no other non-gravitational resistances act, show that the average frictional force must be at least 15 N.

12 A ball, of mass 1 kg, moves in an arc of a vertical circle of radius 1 m by rotating on the end of a light rod. Air resistance can be ignored. Initially the rod hangs vertically. The ball is then given an initial horizontal speed of v m s⁻¹. It travels in a circular arc through an angle θ.

   a Find the gain in the gravitational potential energy of the ball in rising to $\theta = 120°$.
   b Show that the speed of the ball at this position is $\sqrt{v^2 - 30}$ m s⁻¹.
   c In the first case to be considered, $v = 8$. Find the speed of the ball when $\theta = 120°$.
   d In the second case to be considered, the ball comes to rest when $\theta = 120°$. What was its initial speed, v?
   e In the third case to be considered, $v = 3.5$. What is the value of θ when the ball comes to instantaneous rest?
   f In the final case to be considered, the ball is just able to make a complete circle (so its speed at the top of the circular path is 0 m s⁻¹). What was its initial speed, v?

> **EXPLORE 9.1**
>
> In the situation described in Exercise 9A, question 12, suppose that the ball is rotating on a string instead of a rod and that the string breaks when $\theta = 120°$. The ball moves freely under gravity from that point onwards. This means that once the string has broken, the horizontal component of the velocity is constant and the vertical component is subject to a constant acceleration of $10\,\text{m}\,\text{s}^{-2}$ downwards. Use a spreadsheet to investigate where the ball passes through the original vertical circle for different values of the initial speed, v.
>
> You might also investigate the effect of changing the angle at which the string breaks.

9.2 Conservation of energy in a system of conservative forces

A **conservative force** is any force for which the work done by that force in moving a particle between two points is independent of the path taken.

Weight is an example of a conservative force because the work done by the weight depends only on the change in the vertical height between the initial and final positions, and not on the shape of the path taken. Friction and driving force are not conservative forces because the work done depends on the length of the particular path travelled.

When work is done by a conservative force it changes stored potential energy into kinetic energy, with no loss of mechanical energy. All this energy can be recovered again as potential energy by reversing the effect.

In a closed system of conservative forces all energy transfers will be between potential and kinetic energy. You have already seen examples of this in Exercise 9A, questions 5 and 7. In question 7, a boy sat on a sledge at the top of an icy hill. He gently set the sledge in motion. When he reached the bottom of the hill he was moving at $10\,\text{m}\,\text{s}^{-1}$. The initial mechanical energy was all gravitational potential energy, which was then transferred into kinetic energy as the sledge descended the hill. There were no resistance forces so all of the gravitational potential energy was converted into kinetic energy. Taking the bottom of the hill as the zero level for potential energy, the initial potential energy was $10mh$ J, where m kg is the mass of the boy and his sledge and h m is the height of the hill. The initial kinetic energy was 0 J. The final potential energy was 0 J and the final kinetic energy was $50m$ J. As all the potential energy was converted into kinetic energy, this means that the height of the hill is 5 m.

> **KEY POINT 9.3**
>
> A consequence of the work–energy principle is that for a closed system of conservative forces the total mechanical energy, KE + GPE, is constant:
>
> initial KE + initial GPE = KE at any point + GPE at that point = final KE + final GPE
>
> Alternatively, we can think of this as:
>
> loss in GPE = gain in KE (or gain in GPE = loss in KE)
>
> We call this conservation of mechanical energy.

Chapter 9: The work–energy principle and power

WORKED EXAMPLE 9.4

A box of mass m kg is initially at rest. It slides down a smooth slope that is inclined at 30° to the horizontal. Find the speed of the box after sliding a distance of 3 m.

Answer

We can ignore friction and air resistance, so GPE + KE is constant and increase in KE = loss of GPE.

$$\frac{1}{2}m(v^2 - u^2) = mgh$$

$$\frac{1}{2}v^2 = 10h$$

$$v = \sqrt{20h}$$

> There is no mention of resistances so this is a closed system of conservative forces.

> Cancel m and set $u = 0$, $g = 10$.
> The speed is independent of the mass of the box.

After sliding 3 m down the slope:

$h = 3\sin 30° = 1.5$

The box is 1.5 m lower than at the start.

$v = \sqrt{30}$
$= 5.48$

The speed of the box is 5.48 m s^{-1}.

WORKED EXAMPLE 9.5

A ball of mass 0.05 kg is thrown vertically upwards from a height of 1.5 m above the ground. The ball rises through a height of 2 m to reach its maximum height at 3.5 m above the ground. Use the conservation of mechanical energy to find the initial speed of the ball.

Answer

Let the initial speed be u m s^{-1}.

We can ignore friction and air resistance, so GPE + KE is constant.

> There is no mention of resistances so this is a closed system of conservative forces.

Measuring heights from ground level:

initial GPE = $0.05 \times 10 \times 1.5 = 0.75$ J

initial KE = $\frac{1}{2} \times 0.05 \times u^2$ J

Final GPE = $0.05 \times 10 \times 3.5 = 1.75$ J
Final KE = 0 J

Hence, $0.75 + 0.025u^2 = 1.75$
$0.025u^2 = 1$
$u = 6.32$ m s^{-1}

> Alternatively, we could measure heights from the point where the ball was thrown:
>
> initial GPE + initial KE = final GPE + final KE
>
> $$0 + 0.025u^2 = 0.05 \times 10 \times 2 + 0$$
>
> $$u = 6.32 \, \text{m s}^{-1}$$

EXERCISE 9B

1. A parcel of mass 3 kg slides 3.5 m down a smooth slope inclined at 20° to the horizontal. When it reaches the bottom of the slope it has speed 8 m s^{-1}. Find the speed of the parcel at the top of the slope.

2. A waiter drops a plate and it falls 1.43 m to the floor, where it smashes. Find the speed of the plate when it hits the floor.

3. A tennis ball of mass 57 g is hit to give it an initial speed of 180 km h^{-1}. It rises through a height of 1 m. Ignoring air resistance, find:
   a the increase in the gravitational potential energy of the ball
   b the horizontal speed of the ball at the top of its flight.

4. A box slides down a smooth ramp. The height of the ramp is 20 cm and the length of the ramp is 2.5 m. The box starts from rest. What is the speed of the box when it reaches the bottom of the ramp?

5. A ball is launched up a smooth slope that makes an angle 30° to the horizontal. The ball travels a distance 2.5 m up the slope before coming to instantaneous rest. Find the launch speed of the ball.

6. In a pinball machine, ball bearings are fired up a slope inclined at 10° to the horizontal. After travelling 1.2 m the ball bearings reach a curved barrier. Find the maximum initial speed of a ball bearing if it comes to rest before getting to the curved barrier.

7. A boy sits on a sledge at the top of an icy hill. He gently sets the sledge in motion. When he reaches the bottom of the hill he is moving at v m s^{-1}. Assuming that friction is negligible, find an expression for the height of the hill.

8. A diver jumps from a 10 m tall board into a swimming pool.
   a The diver has an initial velocity of u m s^{-1} upwards. Find his speed when he hits the water.
   b What modelling assumptions have been made?

9. A football is kicked from ground level with speed 15 m s^{-1} and rises to a height of 1.45 m. Assume that air resistance is negligible.
   a Find the speed of the ball when it is 1 m above the ground.

   At the top of its flight the ball is travelling horizontally.

   b Explain why the horizontal component of the velocity is constant throughout the motion.
   c Show that the ball was kicked at an angle of 21.0° with the horizontal.

Chapter 9: The work–energy principle and power

 10 A crate of mass M kg sits at the bottom of a smooth slope that is inclined at an angle θ to the horizontal. A light inextensible rope is attached to the crate and passes over a smooth pulley at the top of the slope. The part of the rope between the crate and the pulley is parallel to the slope. The other end of the rope hangs vertically and at the other end there is a ball of mass m kg. The system is released from rest and the ball reaches the ground with speed v m s^{-1} after descending a distance of h m.

   a Find expressions for:

   i the decrease in potential energy for the ball

   ii the increase in kinetic energy for the ball

   iii the increase in mechanical energy for the crate.

   b Use the work–energy principle to show that $v = \sqrt{20h\left(\dfrac{m - M\sin\theta}{m + M}\right)}$.

 11 A piece of sculpture includes a vertical metal circle with radius 2.45 m. A particle of mass 0.2 kg sits at point A on top of the sculpture at the top of the circle (on the outside of the circle). The particle is gently displaced and slides down the circle until it reaches point B, which is level with the centre of the circle. It then falls a further 3.6 m vertically to hit the ground at point C.

   a Use the work–energy principle to find:

   i the speed of the particle when it reaches point B

   ii the speed of the particle when it reaches point C.

   b What modelling assumptions have you made?

   c How do your answers change if the mass of the particle is doubled?

 12 A boy is performing tricks on his skateboard. He skates inside a vertical circle and accelerates until he is moving just fast enough to reach the top of the circle with speed 2 m s^{-1}, using just gravity.

We can model the boy and his skateboard as a particle positioned at his centre of mass, moving in a circle of radius 0.4 m.

   a Find the boy's speed at the bottom of the circle.

   b Find the angle between the upward vertical and the radius from the centre of the circle to the boy when his speed is $\sqrt{10}$ m s^{-1}.

9.3 Conservation of energy in a system with non-conservative forces

A **non-conservative force** is any force for which the work done by that force in moving a particle between two points is different for different paths taken. Driving force, friction and air resistance are examples of non-conservative forces.

When work is done *by* a non-conservative force it converts energy into movement, such as a driving force converting chemical energy from fuel into kinetic energy. The total energy is conserved, but mechanical energy increases.

When work is done *against* a non-conservative force it converts movement into other forms of energy, such as heat energy when friction acts. This energy is 'lost' from the mechanical system. The original situation cannot be recovered by reversing the effect because some mechanical energy has been **dissipated** into non-mechanical energy. The total energy is conserved, but some mechanical energy is lost.

For example, when a box slides across a rough floor and comes to rest because of friction, the kinetic energy is being converted into heat energy (and also some sound energy).

> **DID YOU KNOW?**
>
> Leibniz thought of energy as a 'living force' and believed that the total living force of a body was constant. To account for slowing due to friction, Leibniz said that heat consisted of the random motion of the constituent parts of matter.

Cambridge International AS & A Level Mathematics: Mechanics

WORKED EXAMPLE 9.6

A ball of mass 50 g falls from rest through a height of 80 cm. It hits the ground and rebounds to a height of 30 cm. Find the mechanical energy lost in the motion from the start at height 80 cm to the end at height 30 cm above the ground.

Answer

The initial and final KE are both 0.

The loss of GPE is $0.050 \times 10 \times (0.80 - 0.30) = 0.25$ J .... Convert data to kg and m.

Energy lost $= 0.25$ J .... This will be dissipated as heat and sound.

The system in Worked example 9.6 is not a closed system of conservative forces because the reaction from the ground is a non-conservative force. We cannot recover the original position by reversing the bounce.

WORKED EXAMPLE 9.7

A crate of mass 50 kg slides across a rough horizontal floor. The crate has an initial speed of $3\,\mathrm{m\,s^{-1}}$ and is brought to rest by friction. The distance travelled by the crate is 4 m. Find the coefficient of friction between the floor and the crate.

Answer

The motion is horizontal so there is no change in GPE.

The loss of KE is $\dfrac{1}{2} \times 50 \times (3^2 - 0^2) = 225$ J

Energy lost $= 225$ J .... This will be dissipated as heat and sound.

So work done against friction $= 225$ J .... Use the work–energy principle.

We now need to find the frictional force.

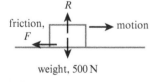

$R = 500$ N .... Resolve vertically.

and $F = \mu R$
$= 500\mu$ .... Crate is moving so friction is limiting.

so work done by friction $= -(-500\mu) \times 4 = 2000\mu$ .... Work done = force × distance

Hence, $2000\mu = 225$

$\mu = 0.1125$

Chapter 9: The work–energy principle and power

WORKED EXAMPLE 9.8

A parcel of mass 3 kg slides 3.5 m down a rough slope inclined at 20° to the horizontal. The coefficient of friction between the parcel and the slope is 0.5. When it reaches the bottom of the slope the parcel has speed 8 m s^{-1}. Use the work–energy principle to find the speed of the parcel at the top of the slope.

Answer

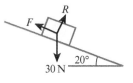

$R = 30 \cos 20 = 28.2$ N — Resolve perpendicular to slope.

$F = 0.5 \times 28.2 = 14.1$ N — Parcel is moving so friction is limiting. $F = \mu R = 0.5 \times R$

Increase in KE + increase in GPE = work done — The only force that does work is friction and work will be done against friction.

Let the speed at the top of the slope be u m s^{-1}.

Increase in KE $= \dfrac{1}{2} \times 3 \times (8^2 - u^2)$

$\qquad = (96 - 1.5u^2)$ J

Increase in GPE $= 3 \times 10 \times (-3.5 \sin 20)$

$\qquad = -35.9$ J

— Parcel slides 3.5 m down slope so vertical drop = 3.5 sin 20 = 1.20 m. The GPE decreases by 35.9 J.

Increase in KE + increase in GPE
$= (96 - 1.5u^2) - 35.9 = (60.1 - 1.5u^2)$ J

The work done against friction $= 14.1 \times 3.5$

$\qquad = 49.3$ J

Hence, $(60.1 - 1.5u^2) = 0 - 49.3$ — Use the work–energy principle.

$1.5u^2 = 109.4$

$u = 8.54$ m s^{-1}

— WD by forces that speed up the parcel = 0 J.

— WD against forces that slow down the parcel = 49.3 J.

Cambridge International AS & A Level Mathematics: Mechanics

WORKED EXAMPLE 9.9

A car of mass 1500 kg (including the driver) is travelling at $64\,\text{km}\,\text{h}^{-1}$ along a level road when the driver sees a ball roll out onto the road and in front of the car. The driver takes 2 s to react and then applies the brakes, using the maximum braking force. The car comes to rest, just missing the ball, after travelling total distance of 80 m from when the driver first saw the ball. Assuming that the wheels lock as soon as the brakes are applied (so the car slides) and that air resistance can be ignored, find the coefficient of friction between the tyres and the road.

Answer

$64\,\text{km}\,\text{h}^{-1} = 17.78\,\text{m}\,\text{s}^{-1}$ — Convert speed to $\text{m}\,\text{s}^{-1}$.

Distance travelled at $17.78\,\text{m}\,\text{s}^{-1} = 2 \times 17.78 = 35.56\,\text{m}$ — Distance travelled while the driver is reacting to seeing the ball.

Distance travelled under braking $= 80 - 35.56 = 44.44\,\text{m}$

Increase in kinetic energy $= 0 - 0.5 \times 1500 \times 17.78^2$

$\qquad = -237\,037\,\text{J}$

So work done against friction $= 237\,037\,\text{J}$ — The car is slowed by the friction between the tyres and the road.

Average frictional force $= \dfrac{237\,037}{44.44}$ — Work done against friction = average frictional force × distance

$\qquad = 5333.33\,\text{N}$

$R = mg$
$\quad = 1500 \times 10$
$\quad = 15\,000\,\text{N}$

$F = \mu R$

$\mu = \dfrac{5333.33}{15\,000}$

$\quad = 0.356$

EXERCISE 9C

1 A helter-skelter ride at a fairground consists of a spiral-shaped slide that people slide down on mats. The top of the slide is 7 m higher than the bottom. The average frictional resistance is 50 N. A boy of mass 60 kg slides down the helter-skelter, starting from rest. At the bottom of the ride the boy has speed $10\,\text{m}\,\text{s}^{-1}$.

   a Find the length of the slide.

   b Work out the amount of mechanical energy that has been lost.

   c What form of energy has most of this loss of mechanical energy been changed into?

2 A box of mass 10 kg slides 3 m down a rough slope inclined at 30° to the horizontal. At the top of the slope the speed of the box is $3.25\,\text{m}\,\text{s}^{-1}$ and at the bottom of the slope the speed of the box is $5\,\text{m}\,\text{s}^{-1}$. Find the coefficient of friction between the box and the slope.

3   A sack of mass 12 kg slides down a ramp, starting from rest at a height of 2 m above the ground. The sack reaches the ground with speed 6 m s^{-1}. Work out the amount of mechanical energy that has been dissipated.

4   A skydiver of mass 80 kg falls 1000 m from rest and then opens his parachute for the remaining 2000 m of his fall. Air resistance is negligible until the parachute opens. The skydiver is travelling at 5 m s^{-1} just before he hits the ground. Find the average resistance force when the skydiver is falling with the parachute open.

PS 5   A tile of mass 1 kg slides 3 m down a roof inclined at 20° to the horizontal. The tile then falls 5 m under gravity to hit the ground with speed 10.5 m s^{-1}. Find the frictional force between the tile and the roof.

6   A model racing car of mass 50 g is released from rest at the top of a downward-sloping track. It travels along the track under the action of gravity. The resistance to the motion of the car is 0.05 N. The car comes to a stop on a horizontal piece of track that is 2 m lower than the top of the track. Find the distance that the car has travelled.

7   A diver jumps from a 10 m board above a swimming pool. The diver has an initial velocity of u m s^{-1} upwards. The horizontal component of the diver's path is negligible. A constant resistance of 0.5 N kg^{-1} acts on the diver. Find an expression for the height of the diver above the pool at the highest point of her dive.

PS 8   A golf ball of mass 45.9 g is hit from a tee with speed 50 m s^{-1}. The ball lands in a pond that is 5 m lower than the tee. When the ball lands in the pond it has travelled along a curved path of length 160 m. The resistance acting on the ball has magnitude 0.3 N.

   a   Find the speed of the ball just before it hits the water.

   The water immediately absorbs 8 J of energy from the ball. The ball then sinks vertically downwards to reach the bottom of the pond. The resistance acting on the ball has magnitude 3 N and the ball just comes to rest as it reaches the bottom of the pond.

   b   Find the depth of the pond.

P 9   A golf ball of mass 45.9 g is hit from a tee with speed 180 km h^{-1}. The ball rises to a height of 20 m, having travelled along a curved path of length 61.875 m. At the highest point of its path the ball is travelling at 144 km h^{-1}.

   a   Find the magnitude of the average resistance force acting on the golf ball.

   The ball travels a further 105.8 m along a curved path to land on the green. The green is 4 m lower than the tee. The average resistance remains unchanged.

   b   Find the speed of the ball just before it lands on the green.

   The ball is travelling vertically when it lands on the green, where it is immediately brought to rest.

   c   Show that the energy absorbed by the green is 28.1 J.

PS 10 Two particles, A and B, are connected by a light inextensible string. Particle A has mass 2 kg and particle B has mass 5 kg. The string passes over a pulley and hangs vertically with particle A and particle B on each side of the pulley. The pulley, however, is not smooth and 10 J of energy is dissipated for each rotation of the pulley. The system is released from rest, and the particles reach a speed of 0.2 m s^{-1} after each moving 1.6 m.

   a   Work out how many rotations the pulley has made.

   b   If the string passes over the pulley without slipping, work out the radius of the pulley.

11 A woman of weight 54 N skis from point X to point Y. The distance from point X to point Y is 16.2 m. Point Y is 3 m lower than point X. At point X she has speed 1 m s^{-1} and at point Y she has speed 7 m s^{-1}.

   a   Use the work–energy principle to work out the average resistance force that acts on the woman.

   b   Give an expression for the average resistance force if, instead, her speed at point Y is v m s^{-1}.

12 A piece of sculpture includes a vertical metal circle with radius 2.45 m. A particle of mass 0.2 kg sits at point A on top of the sculpture at the top of the circle (on the outside of the circle). The particle is gently displaced and slides down the circle until it reaches point B, which is level with the centre of the circle. It then falls a further 3.6 m vertically to hit the ground at point C. When the particle reaches point C it has speed $10\,\text{m s}^{-1}$. Air resistance can be ignored.

  a Work out how much mechanical energy has been lost by the particle in travelling from A to C.

  b Show that the average frictional force between the surface and the particle is 0.546 N.

  c It is claimed that the coefficient of friction between the surface and the particle is 0.273. Explain how this value has been calculated and why it is too small.

9.4 Power

Energy can be put into a system by an engine converting fuel (chemical energy) into a driving force. The work done by the engine is given by:

$$\text{work done} = \text{force} \times \text{displacement}$$
$$= Fs$$

Power is the rate of doing work, so the average power generated by an engine is given by:

$$\text{average power} = \frac{\text{work done by the engine}}{\text{time taken}}$$
$$= \frac{Fs}{t}$$

It always takes the same amount of work to make a car speed up from 0 to $100\,\text{km h}^{-1}$, but a more powerful engine will get the car to $100\,\text{km h}^{-1}$ more quickly than a less powerful one.

Over a very small time interval, δt, the driving force F is constant and the rate of doing work is given by:

$$\text{power} = F\frac{\delta s}{\delta t}$$

As δt gets smaller, $\frac{\delta s}{\delta t}$ approaches the limit $\frac{ds}{dt}$ and we get:

$$\text{power} = F\frac{ds}{dt} = Fv$$

> **REWIND**
>
> Recall, from Chapter 6, that when we differentiate displacement with respect to time we get velocity, and when we differentiate distance with respect to time we get speed.

> **KEY POINT 9.4**
>
> The rate at which an engine works is called the power of the engine.
>
> $$\text{Power} = \text{rate of doing work} = Fv$$
>
> where F, the driving force, is constant.

Power is measured in J s^{-1} or watts (W). We often use units of 1000 watts (kW).

Power is a scalar quantity. Strictly speaking, this involves a product of vectors, but as we are usually only concerned with motion in one direction along a line, we can say that

$$\text{power} = \text{force} \times \text{speed}$$

Chapter 9: The work–energy principle and power

MODELLING ASSUMPTIONS

When an engine converts energy from fuel into other forms of energy, some energy is lost in the form of heat and sound. We will ignore this and assume that the stated power of an engine is the measure of the rate of energy conversion to mechanical energy by the engine and that no energy is lost. However, energy losses do need to be considered by manufacturers of machines so they can try to minimise energy losses and improve efficiency.

WORKED EXAMPLE 9.10

A car of mass $1500\,\text{kg}$ is being driven along a level road. It accelerates from $0\,\text{km}\,\text{h}^{-1}$ to $100\,\text{km}\,\text{h}^{-1}$ in $10\,\text{s}$. Air resistance and friction may be ignored. Find the average power generated by the engine.

Answer

$100\,\text{km}\,\text{h}^{-1} = 27.8\,\text{m}\,\text{s}^{-1}$ Convert speed to $\text{m}\,\text{s}^{-1}$.

Work done $= 0.5 \times 1500 \times (27.8^2 - 0^2) = 578\,704\,\text{J}$ Increase in KE = WD by engine

Average power $= 578\,704 \div 10$ Average power = WD ÷ time taken
$= 57870.4\,\text{W}$
$= 57.9\,\text{kW}$ (to 3 significant figures)

For a given power, the driving force generated will be greater at lower speeds and smaller at higher speeds.

For example, a car pulling off from stationary under maximum power has a low speed and so has a greater driving force than when it is moving more quickly. This means that the resultant force is greater and so, using Newton's second law, the acceleration is greater and the car speeds up quickly.

As the speed increases, the driving force (under maximum power) decreases and so the resultant force decreases and, hence, the acceleration decreases. If the maximum power is maintained, the acceleration will eventually become 0 and the driving force is balanced by the resistances. At this point, the car is moving at its maximum speed and this cannot be increased further.

We can use the maximum power output of an engine to find the maximum speed that it can generate. At the maximum speed the acceleration is $0\,\text{m}\,\text{s}^{-2}$ and, hence, the resultant force is $0\,\text{N}$.

TIP

It is important to be very careful to use the *resultant* force (or net force) in Newton's second law but only the *driving* force in the power equation. You may find it helpful to denote the driving force by D rather than F.

WORKED EXAMPLE 9.11

A car of mass $1500\,\text{kg}$ has an engine that has a maximum power output of $200\,\text{kW}$. The resistance to motion is typically $5000\,\text{N}$. Find the maximum speed that the car can achieve on a level road (ignoring legal speed restrictions).

Answer

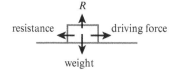

A diagram is helpful.

Resistance = 5000 N

Resultant force = driving force − 5000
At the maximum speed, acceleration = 0.   ······· Use Newton's second law.
Resultant force = 0
Driving force = 5000 N

Power = driving force × speed   ······· Note: we have not used the mass of the car,
= 5000v so the maximum speed is the same for any
                                                    engine that has this maximum power and this
$v = \dfrac{200\,000}{5000} = 40\,\text{m s}^{-1}$ resistance to motion.

WORKED EXAMPLE 9.12

A car of mass 1500 kg has an engine that has a maximum power output of 200 kW. The resistance to motion is typically 5000 N. Find the instantaneous acceleration of the car when the engine is working at its maximum power and the car is travelling at $20\,\text{m s}^{-1}$.

Answer

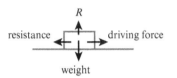

Power = driving force × speed
Driving force = 200 000 ÷ 20
            = 10 000 N
Resultant force = driving force − resistance
              = 10000 − 5000
              = 5000 N
Resultant force = mass × acceleration   ······· Use Newton's second law.
Instantaneous acceleration = 5000 ÷ 1500
                         = $3.33\,\text{m s}^{-2}$

A diagram is helpful.

Resistance = 5000 N

WORKED EXAMPLE 9.13

A car of mass 1500 kg has an engine that has a maximum power output of 200 kW. The resistance to motion is typically 5000 N. Find the instantaneous acceleration of the car when the engine is working at its maximum power and the car is travelling at $20\,\text{m s}^{-1}$ up a hill that is inclined at $\sin^{-1} 0.28$ to the horizontal.

Answer

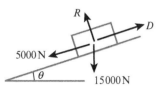

sin θ = 0.28

A diagram is helpful.

Resistance = 5000 N

Component of weight down slope
= 15 000 sin θ = 4200 N

Power = driving force D × speed
$D = 200\,000 \div 20$
$= 10\,000$ N
Resultant force (up slope) $= 10\,000 - 5000 - 4200$
$= 800$ N

Resultant force = mass × acceleration ⋯⋯⋯⋯ Use Newton's second law.
Instantaneous acceleration $= 800 \div 1500$
$= 0.533\,\text{m s}^{-2}$

EXERCISE 9D

1 A car of mass 2000 kg travels in a straight line on a horizontal road. The car accelerates from $10\,\text{m s}^{-1}$ to $20\,\text{m s}^{-1}$ in 8 s. Assume that resistance can be ignored.

   a Use the work–energy principle to find the work done by the driving force.

   b Find the average power generated by the engine.

2 The engine of a 5 tonne truck has a power output of 400 kW. The truck is travelling in a straight line on a horizontal road. The resistance to motion is 20 000 N. Find the maximum speed the truck could achieve.

3 The maximum power of a boat engine is 140 kW. The boat is subject to a resistance force of 10 000 N. Find the maximum speed the boat can achieve when travelling in a straight line.

4 A model train of mass 200 g is moving in a straight line on a level track. The train accelerates from $2\,\text{m s}^{-1}$ to $8\,\text{m s}^{-1}$ in 3 s. Find the average power generated by the engine.

5 The maximum power of a boat engine is 120 kW. The maximum speed the boat can achieve is $10\,\text{m s}^{-1}$. Find the resistance force acting on the boat when it is travelling at its maximum speed and the engine is working at maximum power.

6 A car of mass 1800 kg is being driven, at a constant $15\,\text{m s}^{-1}$, against a resistance of 2000 N, up a hill inclined at 10° to the horizontal. Find the rate at which the engine is working.

7 A car of mass 1600 kg is being driven up a hill inclined at 10° to the horizontal. The car has an initial speed of $10\,\text{m s}^{-1}$ and a final speed of $12\,\text{m s}^{-1}$ after 60 s. Air resistance and friction may be ignored. Assuming a constant acceleration, find the power being exerted by the engine when the car is travelling at $11\,\text{m s}^{-1}$.

8 A car of mass 1200 kg accelerates up a hill against a resistance of 263 N. At a certain point on the hill the road is inclined at 8° to the horizontal. The engine is working at 75 kW and the car is travelling at $25\,\text{m s}^{-1}$. Find the acceleration of the car at this instant.

9 A small van of mass 1600 kg accelerates from rest in a straight line along a horizontal road. The resistance from friction and air resistance is 2400 N throughout the motion. The engine works at a constant rate of 48 kW.

   a Write down an expression for the driving force when the van is travelling at $v\,\text{m s}^{-1}$.

   b Write down an expression for the acceleration of the van when it is travelling at $v\,\text{m s}^{-1}$.

   c Explain why the power cannot be constant.

10 A powerboat of mass 500 kg travels in a straight line at its maximum velocity against a resistance of 15 N. The engine of the powerboat has a maximum power output of 750 W.

  a Find the maximum velocity.

  In different weather conditions the same powerboat has a maximum velocity of only $25\,\mathrm{m\,s^{-1}}$.

  b State what has changed in the model and give the new value of this quantity.

11 A van of mass m kg moves up a hill that is inclined at 2° to the horizontal. The engine works at a constant rate of 30 kW and the resistance (from friction and air resistance) is a constant 400 N. When the van is travelling at $20\,\mathrm{m\,s^{-1}}$ it has acceleration $0.1\,\mathrm{m\,s^{-2}}$. Find the value of m.

12 Car A, of mass 1250 kg, is travelling along a straight horizontal road at speed $10\,\mathrm{m\,s^{-1}}$. The engine works at a constant rate of 25 kW and the resistance is a constant 500 N. After 5 s the speed of the car has increased to $v\,\mathrm{m\,s^{-1}}$.

  a Use the work–energy principle to find the amount of energy that is dissipated and, hence, find the distance travelled in the 5 s.

  b Find an expression for the acceleration at time 5 s as a function of v and show that the acceleration is not constant.

  Car B travels along the same road, starting with speed $10\,\mathrm{m\,s^{-1}}$ and accelerating at a constant rate for 5 s. After 5 s the two cars have the same speed and also have the same acceleration as one another.

  c Show that v must satisfy the equation $v^2 - 8v = 100$ and, hence, find the speed of the cars at the end of the 5 s.

Checklist of learning and understanding

- The work–energy principle states that for *any* motion:

  increase in kinetic energy = total work done by all forces

  $$\frac{1}{2}mv^2 - \frac{1}{2}mu^2 = \sum Fs$$

  where the 'total work done' is the sum of the work done by forces with a component in the direction of motion (forces that speed up the motion) minus the work done against forces with a component in the direction opposing the motion (forces that slow down the motion).

- The work–energy principle can also be written as:

  increase in mechanical energy = total work done by forces that help to speed the body up
  − total work done by forces that help to slow the body down

  (In both cases 'forces' excludes the weight of the body.)

- Kinetic energy and potential energy are types of mechanical energy. Other forms of energy (heat, light, sound, chemical, electrical, nuclear etc.) are non-mechanical energy.

- A consequence of the work–energy principle is that for a system of conservative forces the total mechanical energy is constant. We call this conservation of mechanical energy.

- Power is measured in Watts. The power of an engine, in watts, is the rate at which that engine can work:

  power = work done ÷ time taken

- The power of an engine is also the product of the driving force of the engine and the velocity in the direction of the driving force:

  power = Fv, where F is the driving force of the engine.

Chapter 9: The work–energy principle and power

END-OF-CHAPTER REVIEW EXERCISE 9

1. A car of mass 1250 kg is travelling along a straight horizontal road with its engine working at a constant rate of P W. The resistance to the car's motion is constant and equal to R N. When the speed of the car is $19\,\text{m s}^{-1}$ its acceleration is $0.6\,\text{m s}^{-2}$, and when the speed of the car is $30\,\text{m s}^{-1}$ its acceleration is $0.16\,\text{m s}^{-2}$. Find the values of P and R. [6]

   Cambridge International AS & A Level Mathematics 9709 Paper 43 Q2 June 2011

2. Particle X, of mass 2 kg, and particle Y, of mass m kg, are attached to the ends of a light inextensible string of length 4.8 m. The string passes over a fixed small smooth pulley and hangs vertically either side of the pulley. Particle X is held at ground level, 3 m below the pulley. Particle X is released and rises while particle Y descends to the ground.

   a Find an expression, in terms of m, for the tension in the string while both particles are moving. [2]

   b Use the work–energy principle to find how close particle X gets to the pulley in the subsequent motion. [2]

3. A van of mass 1500 kg starts from rest. It is driven in a straight line up a slope inclined at angle α to the horizontal, where $\sin\alpha = \dfrac{1}{10}$. The driving force of the engine is 2000 N and the non-gravitational resistances total 350 N throughout the motion. The speed of the van is $v\,\text{m s}^{-1}$ when it has travelled x m from the start. Use the work–energy principle to find v in terms of x. [6]

4. A car of mass 1000 kg travels in a straight line up a slope inclined at angle α to the horizontal, where $\sin\alpha = 0.05$. The non-gravitational resistances are 200 N throughout the motion.

   a When the power produced by the engine is 50 kW, the car is accelerating at $1.2\,\text{m s}^{-2}$. Find the speed of the car at this instant. [4]

   b What would happen to the speed if the mass of the car increased? [1]

   c What would happen to the speed if the power produced by the engine decreased? [1]

5. A truck of mass 3000 kg starts from rest. It is driven in a straight line up a slope inclined at angle α to the horizontal, where $\sin\alpha = 0.08$. The driving force of the engine is 7000 N and the non-gravitational resistances total 4000 N throughout the motion. The speed of the truck is $v\,\text{m s}^{-1}$ when it has travelled x m from the start. Find the value of k for which $x = kv^2$. [6]

6. A car of mass 1200 kg is driven along a straight horizontal road against a resistance of 5000 N. The engine has a maximum power output of 100 kW.

   a Find the maximum speed the car can reach. [2]

   b Find the power being used when the car is travelling at a speed of $15\,\text{m s}^{-1}$ and accelerating at $-2\,\text{m s}^{-2}$. [4]

7. A box of mass 2 kg is pulled up a rough slope by a rope. The rope passes over a smooth pulley and is attached, at the other end, to a block of mass 4 kg with that end of the rope hanging vertically. The slope is inclined at 20° to the horizontal and the coefficient of friction between the slope and the box is 0.3. The system is released from rest. Use the work–energy principle to find the speed of the box when it has moved 1 m up the slope. [6]

8. A car of mass 1600 kg travels down a straight road inclined at an angle θ to the horizontal. The power produced by the engine is 24 kW and the non-gravitational resistance is a constant 1600 N.

   a Find the driving force when the car is travelling at a constant $20\,\text{m s}^{-1}$. [2]

   b Find the value of $\sin\theta$. [4]

9 A lorry of mass 14 000 kg moves along a road starting from rest at a point O. It reaches a point A, and then continues to a point B which it reaches with a speed of $24\,\mathrm{m\,s^{-1}}$. The part OA of the road is straight and horizontal and has length 400 m. The part AB of the road is straight and is inclined downwards at an angle of $\theta°$ to the horizontal and has length 300 m.

   i   For the motion from O to B, find the gain in kinetic energy of the lorry and express its loss in potential energy in terms of θ. [3]

   The resistance to the motion of the lorry is 4800 N and the work done by the driving force of the lorry from O to B is 5000 kJ.

   ii Find the value of θ. [3]

   Cambridge International AS & A Level Mathematics 9709 Paper 41 Q4 June 2015

10 A block of mass 25 kg is dragged across a rough horizontal floor, using a rope that makes an angle of 30° with the floor. The coefficient of friction between the floor and the block is 0.25. The tension in the rope is T N and air resistance can be ignored. After travelling a distance of 5 m, the speed of the box has increased by $2\,\mathrm{m\,s^{-1}}$.

   a   Find the work done against friction, in terms of T. [3]

   b   Use the work–energy principle to find, in terms of T, the average of the initial and final speeds. [4]

11 A light inextensible rope has a block A of mass 5 kg attached at one end, and a block B of mass 16 kg attached at the other end. The rope passes over a smooth pulley which is fixed at the top of a rough plane inclined at an angle of 30° to the horizontal. Block A is held at rest at the bottom of the plane and block B hangs below the pulley (see diagram). The coefficient of friction between A and the plane is $\dfrac{1}{\sqrt{3}}$. Block A is released from rest and the system starts to move. When each of the blocks has moved a distance of x m each has speed $v\,\mathrm{m\,s^{-1}}$.

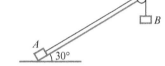

   i   Write down the gain in kinetic energy of the system in terms of v. [1]

   ii Find, in terms of x,

       a   the loss of gravitational potential energy of the system, [2]

       b   the work done against the frictional force. [3]

   iii Show that $21v^2 = 220x$. [2]

   Cambridge International AS & A Level Mathematics 9709 Paper 42 Q5 June 2014

12 Particle W, of mass 3 kg, and particle X, of mass 5 kg, are attached to the ends of a light, inextensible string of length 4 m. The string passes over a small smooth pulley fixed at the top of a fixed triangular wedge, ABC. The angles BAC and BCA are each 45° and the side AC is fixed to horizontal ground. The distance from A to C is $3\sqrt{2}$ m.

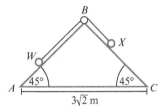

   Surface AB is smooth and surface BC is rough, with coefficient of friction $\dfrac{1}{\sqrt{8}}$.

   Particle W is held at the bottom of the slope AB and is then gently released.

   a   Find the work done against friction when particle X moves a distance x m. [3]

   b   Find the change in the total potential energy when particle X moves a distance x m. [3]

   c   Use the work–energy principle to find the speed of the particles when particle X reaches the ground at C. [2]

   d   Explain why the work done by the tension does not need to be included in the work–energy calculation. [1]

CROSS-TOPIC REVIEW EXERCISE 3

1. A ball of mass 0.5 kg is dropped from a height of 0.45 m and bounces to a height of 0.2 m. Find the change in momentum of the ball during the bounce. [5]

2. A box of mass 3 kg is pulled 10 m from rest up a smooth slope, which is at an angle of 8° to the horizontal. The box is pulled by a string with constant tension, parallel to the line of greatest slope, for 6 s. Find the work done by the tension in the string. [5]

3. Object A has mass 3 kg and is moving with velocity $10\,\text{m s}^{-1}$ towards object B, which has mass 6 kg and is stationary. After they collide, object A bounces back with speed $2\,\text{m s}^{-1}$. Object B then collides with object C, which has mass 17 kg and is stationary. Object C moves at $3\,\text{m s}^{-1}$ after this collision. Deduce whether or not there will be a third collision and explain your reasoning. [5]

4. A cyclist and his cycle have a combined mass of 80 kg. He works at a rate of 600 W while cycling along a straight horizontal road. There is a constant resistance of R N.

   a Given the cyclist has a maximum velocity of $12\,\text{m s}^{-1}$, find the value of R. [2]

   b Find the speed of the cyclist when he is accelerating at $0.625\,\text{m s}^{-2}$. [3]

5. A particle of mass 3 kg is projected with speed $4\,\text{m s}^{-1}$ towards a stationary particle of mass 5 kg.

   a The particles coalesce on impact. Find the speed at which the particles move after the collision. [2]

   b The coalesced particles then move towards a particle of mass 2 kg. After the collision the coalesced particles remain stationary and the 2 kg particle moves with speed $3\,\text{m s}^{-1}$. Find the speed and direction of motion of the 2 kg particle before the collision. [3]

6. A car of mass 1200 kg travels along a straight horizontal road, starting at a point A. The resistance to motion of the car is 800 N.

   a The car travels from A to a point B at a constant speed in 12 s. The power of the engine is 20 kW. Find the distance AB. [3]

   b The car travels from B to a point C with an increased power of 24 kW, reaching C with a speed of $28\,\text{m s}^{-1}$ after 37 s. Find the distance BC. [3]

7. The diagram shows a vertical cross-section, $ABCD$, of a fixed surface. AB and CD are smooth curves and BC is a rough horizontal surface. A is at a vertical height 3.2 m above BC. A particle of mass 2 kg is released from A and travels along the surface to D.

   a Find the speed of the particle at B. [2]

   b Given that the particle reaches C with a speed of $4\,\text{m s}^{-1}$, find the work done against the resistance to motion as the particle moves from B to C. [2]

   c The particle reaches the point D. Find the maximum vertical height of D above BC. [3]

8. A car of mass 2000 kg climbs a straight hill, ABC, which makes an angle θ with the horizontal, where $\sin\theta = \dfrac{1}{16}$. For the motion from A to B, the work done by the car's engine is 256 kJ and the resistance to motion is R N. The length of AB is 200 m. The speed of the car is $20\,\text{m s}^{-1}$ at A and $16\,\text{m s}^{-1}$ at B.

   a Find the value of R. [4]

   b From B to C, the work done by the engine is 388 kJ. The resistance to motion remains the same as that between A and B. The speed of the car at C is $12\,\text{m s}^{-1}$. Find the distance BC. [3]

9 A particle, P, of mass 4 kg is projected with speed 9 m s^{-1} along rough horizontal ground. The coefficient of friction between P and the ground is 0.4. After 7 m it strikes a stationary particle, Q, of mass 3 kg. The coefficient of friction between Q and the ground is also 0.4. Q comes to rest after 2 m.

   a Show that the speed of P immediately before the collision is 5 m s^{-1} and find the speed of Q immediately after the collision. [5]

   b Find the distance between Q and P when both have come to rest. [3]

10 A lorry of mass 24 000 kg is travelling up a hill which is inclined at 3° to the horizontal. The power developed by the lorry's engine is constant, and there is a constant resistance to motion of 3200 N.

   i When the speed of the lorry is 25 m s^{-1}, its acceleration is 0.2 m s^{-2}. Find the power developed by the lorry's engine. [4]

   ii Find the steady speed at which the lorry moves up the hill if the power is 500 kW and the resistance remains 3200 N. [5]

   Cambridge International AS & A Level Mathematics 9709 Paper 41 Q3 November 2015

11 A box of mass 25 kg is pulled, at a constant speed, a distance of 36 m up a rough plane inclined at an angle of 20° to the horizontal. The box moves up a line of greatest slope against a constant frictional force of 40 N. The force pulling the box is parallel to the line of greatest slope. Find

   i the work done against friction, [1]

   ii the change in gravitational potential energy of the box, [2]

   iii the work done by the pulling force. [2]

   Cambridge International AS & A Level Mathematics 9709 Paper 41 Q2 June 2016

12 A car of mass 1000 kg is moving along a straight horizontal road against resistances of total magnitude 300 N.

   i Find, in kW, the rate at which the engine of the car is working when the car has a constant speed of 40 m s^{-1}. [3]

   ii Find the acceleration of the car when its speed is 25 m s^{-1} and the engine is working at 90% of the power found in part i. [3]

   Cambridge International AS & A Level Mathematics 9709 Paper 41 Q3 June 2016

13 A block of mass 25 kg is pulled along horizontal ground by a force of magnitude 50 N inclined at 10° above the horizontal. The block starts from rest and travels a distance of 20 m. There is a constant resistance force of magnitude 30 N opposing motion.

   i Find the work done by the pulling force. [2]

   ii Use an energy method to find the speed of the block when it has moved a distance of 20 m. [2]

   iii Find the greatest power exerted by the 50 N force. [2]

After the block has travelled the 20 m, it comes to a plane inclined at 5° to the horizontal. The force of 50 N is now inclined at an angle of 10° to the plane and pulls the block directly up the plane (see diagram). The resistance force remains 30 N.

   iv Find the time it takes for the block to come to rest from the instant when it reaches the foot of the inclined plane. [4]

   Cambridge International AS & A Level Mathematics 9709 Paper 41 Q6 November 2016

PRACTICE EXAM-STYLE PAPER

Time allowed is 1 hour 15 minutes (50 marks).

Answer all the questions.

Give non-exact numerical answers correct to 3 significant figures, or 1 decimal place in the case of angles in degrees, unless a different level of accuracy is specified in the question.

Where a numerical value for the acceleration due to gravity is needed, use $10\,\text{m s}^{-2}$.

The use of an electronic calculator is expected, where appropriate.

You are reminded of the need for clear presentation in your answers.

1. Four horizontal forces act at a point. The forces have magnitudes F N, 5 N, 4 N and 10 N. The F N force acts at an angle of 90° to the 5 N force and at an angle of 120° to the 10 N force. The 4 N force acts at an angle α to the 5 N force, as shown in the diagram. The forces are in equilibrium. Show that $\alpha = 23.8°$ and find the value of F. [4]

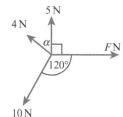

2. A car has a maximum power output of 60 kW. The car is driven at its maximum power in a straight line on a horizontal road against a constant resistance. The car travels 200 m at a constant speed of $32\,\text{m s}^{-1}$.

   a Find the resistance. [2]

   b Find the work done by the engine. [2]

3. Two balls are travelling directly towards one another in a straight line. The first ball has mass 1.2 kg and is initially moving at $3\,\text{m s}^{-1}$; the second ball is initially moving at $5\,\text{m s}^{-1}$. The balls hit each other and after the impact each ball has reversed its direction of travel and is moving at half its original speed.

   a Find the mass of the second ball. [3]

   b Find the loss of kinetic energy in the impact. [2]

4. A girl roller-skates in a straight line along a horizontal track. She starts from rest and accelerates at a constant rate for 20 s, during which time she covers a distance of 100 m. She travels at a constant speed for the next 10 s and then slows at a constant rate for 5 s until she stops.

   a Find the total distance that the girl skates. [4]

   On another occasion the girl skates along the same track, starting from rest, with the same acceleration as before. This time she accelerates for only 10 s before travelling at a constant speed.

   b How long does she take to travel 100 m? [2]

5 A man runs along a straight track, starting from rest. He accelerates so that his velocity, v_1 m s^{-1}, is given by $v_1 = \frac{1}{3}t^2 - \frac{1}{27}t^3$, where t is the time, in seconds, from when he starts to run. The man runs until his acceleration $a_1 = 0$ and then runs at constant speed V for 50 m. He then accelerates again, with acceleration a_2 m s^{-2}, given by $a_2 = 0.048T^2 - 0.008T^3$, where T is the time, in seconds, from when he starts the second acceleration phase. The man runs until he comes to rest and then stops.

   a   Find V. [4]

   b   Show that the man comes to rest when $T = 10$. [3]

   c   Find the total time for which the man runs. [2]

6 Particles A and B, of masses 0.4 kg and 0.1 kg, respectively, are attached to the ends of a light inextensible string. Particle A sits on a rough horizontal table and the string passes over a small smooth pulley at the edge of the table.

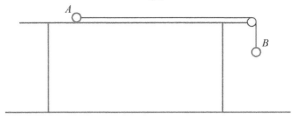

The system is released from rest and particle B descends 1 m in 2 s.

   a   Calculate the frictional force acting on particle A. [5]

Particle B is now on the floor. Particle A continues until it comes to rest, without having reached the pulley.

   b   Find the total distance travelled by particle A. [5]

7 A crate of mass 29 kg slides down a slope inclined at an angle θ to the horizontal, where $\sin\theta = \frac{17}{145}$. For the first part of the slope the coefficient of friction between the slope and the crate is $\frac{1}{25}$.

   a   Find the acceleration of the crate down the slope on this part of the slope. [5]

When the crate is moving at 0.3 m s^{-1} down the slope, the surface of the slope changes, although the angle of the slope is unchanged. After travelling 0.58 m on this second part of the slope the crate is moving at 0.1 m s^{-1}.

   b   Find the loss in the kinetic energy of the crate. [2]

   c   Find the loss in the potential energy of the crate. [2]

   d   Find the average resistance force on the crate while it is travelling on this second part of the slope. [3]

Answers

1 Velocity and acceleration

Prerequisite knowledge
1.   a $x = -3$ or $x = 5$ b $x = 1$ or $x = -\frac{3}{2}$
    c $x = -0.907$ or $x = 2.57$
2.   a $x = 1$ and $y = 2$ b $x = 1$ and $y = 3$

Exercise 1A
1. $8\,\text{m}\,\text{s}^{-1}$
2. $63\,\text{m}$
3.   a $6\,\text{s}$
    b   The cheetah can instantly reach that speed. The gazelle remains stationary.
4.   8 minutes 20 seconds
5. $2.94\,\text{s}$
6.   a $6.3\,\text{m}\,\text{s}^{-1}$
    b   The speeds are average speeds or the runner instantaneously changes speed between sections.
7.   a $45\,\text{m}$ b $3\,\text{m}\,\text{s}^{-1}$
    c $5\,\text{m}\,\text{s}^{-1}$
8. $12.5\,\text{m}\,\text{s}^{-1}$
9. $0.09\,\text{s}$
10. $80\,\text{s}$
11. $15\,840\,\text{m}$
12. $1.014\,\text{m}$ in $0.78\,\text{s}$
13.   a   Proof   b   Proof
14.   a   Proof   b   Proof

Exercise 1B
1. $2\,\text{m}\,\text{s}^{-2}$
2. $2.5\,\text{m}\,\text{s}^{-2}$
3. $1.5\,\text{s}$
4. $19\,\text{m}\,\text{s}^{-1}$
5. $3\,\text{m}\,\text{s}^{-1}$
6. $14\,\text{m}\,\text{s}^{-1}$
7. $48\,\text{m}$
8.   a $0.6\,\text{m}\,\text{s}^{-2}$
    b   The sprinter can maintain a constant acceleration and we are ignoring the shape of the sprinter's body and the different positions it takes when running, by considering the sprinter having a single position at any point in time.
9. $3\,\text{m}\,\text{s}^{-2}$
10. $8\,\text{m}\,\text{s}^{-1}$
11.   He can pedal because doing nothing he will arrive at the bend with velocity $10.8\,\text{m}\,\text{s}^{-1}$.

Exercise 1C
1.   a $s = 32\,\text{m}$ b $s = 72\,\text{m}$
    c $a = 2\,\text{m}\,\text{s}^{-2}$ d $a = 3\,\text{m}\,\text{s}^{-2}$
    e $a = 4\,\text{m}\,\text{s}^{-2}$ f $u = 3\,\text{m}\,\text{s}^{-1}$
    g $v = 1\,\text{m}\,\text{s}^{-1}$ h $s = 14\,\text{m}$
2.   a $t = 4\,\text{s}$ b $t = 6\,\text{s}$
    c $t = 4\,\text{s}$
3. $v = -1\,\text{m}\,\text{s}^{-1}$
4. $u = 7\,\text{m}\,\text{s}^{-1}$
5.   a $v = 9\,\text{m}\,\text{s}^{-1}$
    b   Positive acceleration means v must be larger than u.
6. $50\,\text{m}$
7. $4.5\,\text{m}\,\text{s}^{-2}$
8. $800\,\text{m}$
9. $5\,\text{m}\,\text{s}^{-2}$
10.   a $10\,\text{m}\,\text{s}^{-1}$
    b   The deceleration is constant.
11. $0.4\,\text{m}$ past the target
12.   No, the ball stops $0.4\,\text{m}$ short of the hole.

13 The car cannot brake safely in time, but can accelerate to get past the lights before they turn red, so must accelerate.

14 a Proof   b Proof

15 Proof

16 Proof

Exercise 1D

1 a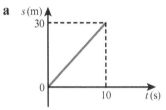

b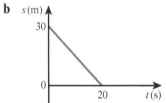

c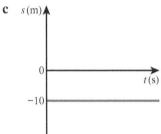

d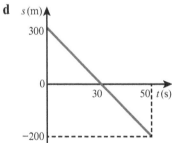

2 a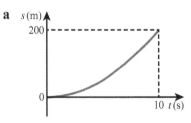

b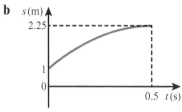

c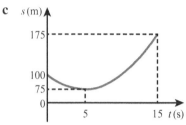

d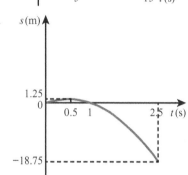

3 a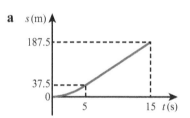

b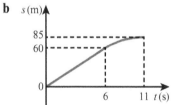

c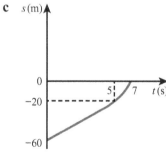

d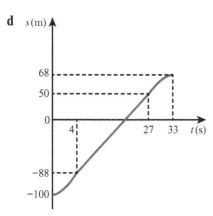

4 $s = 16t - 60$ so $t = 3.75\,\text{s}$

5 a $p = -0.5$, $q = 10$ and $r = 0$
  b $s = -50 + 10t - \frac{1}{2}t^2$
  So $a = -1\,\text{m s}^{-2}$ and $u = 10\,\text{m s}^{-1}$

6 a s(m) — graph showing two lines crossing, one starting at 0 and one starting at 3000

  b $s = 30t$ and $s = 3000 - 20t$
  c They meet at $t = 60\,\text{s}$ at a distance of $1800\,\text{m}$ from junction 1.

7 a s(m) — graph showing two lines crossing, one starting at 0 passing near 40, and one starting at 5000

  b $t = 140\,\text{s}$ and $s = 3500\,\text{m}$

8 $s = 8t$ and $s = 25 + 10(t - 5)$
  $12.5\,\text{s}$

9 a $115\,\text{s}$
  b Rowing boats can travel at constant speed (in reality they tend to increase speed with the strokes and decrease speed between strokes).

10 $0.02\,\text{m s}^{-2}$

11 $60\,\text{m}$

12 $50\,\text{s}$

13 $1.2\,\text{m}$

14 $6\,\text{s}$

15 $h = \dfrac{gt_1t_2}{2}$

Exercise 1E

1 **a**, **b**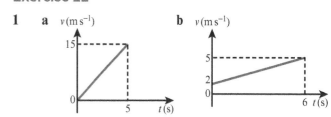

  c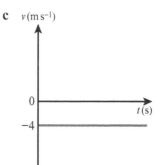

  d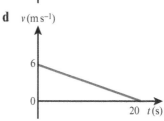

2 a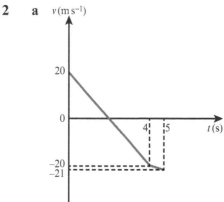

  b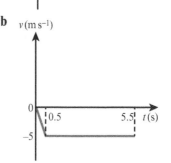

c $v\,(\mathrm{m\,s^{-1}})$

d $v\,(\mathrm{m\,s^{-1}})$

3 160 m
4 6.25 m
5 96 m
6 1260 m
7 22.5 s
8 11.25 s
9 12
10 a 8
   b The boat accelerates at a constant rate and moves instantaneously to constant speed at the change between the two stages.
11 1.19
12 a 62 m   b 23.1 s
13 a He leads by 69 m.   b 64.7 s
14 a $v\,(\mathrm{m\,s^{-1}})$

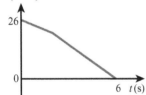

   b $v = 26 - 3t$ and $v = -5(t-6)$ c 2 s
15 a The graph is a triangle and area under graph is $s = \frac{1}{2} tv$ independent of gradients of lines.
   b The graph is a trapezium and area under graph is $s = \frac{1}{2}(t+T)\,v$, independent of gradients of lines.

Exercise 1F

1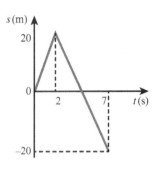

2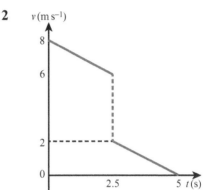

3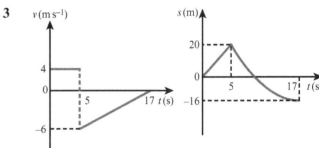

4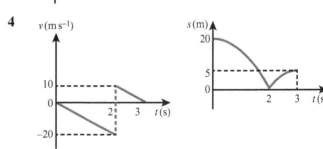

5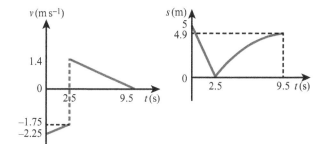

6 a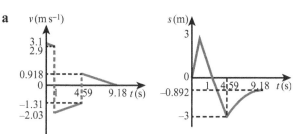

  b The ball is modelled as a particle so has no width (otherwise the distance the ball travels between the cushions would be less than 6m) and the change in velocity is instantaneous.

7 Proof

End-of-chapter review exercise 1
1 a 25 m   b 5.67 s
2 a 112.5 m   b 23 s
3 a 0.008 m s^{-2} b 0.075 m s^{-2}
4 a 6.25 m s^{-2} b 46.9 m s^{-2}
5 s (m)

6 5 m s^{-1}
7 a 4.00 s
   b When the footballer kicks the ball it instantaneously starts moving at 4 m s^{-1}.
8 a 135 m   b 11.6 s
9 a Proof
   b The closest the lion gets is 0.5 m away at $t = 10$ s.
10 a Proof   b 6.5 s
   c 320 m
11 55 m
12 640 m
13 Proof
14 $\dfrac{2u^2 s - 2usv - as^2}{2(u-v)^2}$
15 i   0.02 m s^{-2}, -0.21 m s^{-2}
   ii 42.5 m   iii 86.5 m
16 i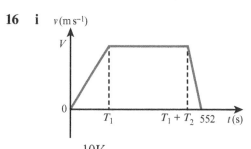

   $T_1 = \dfrac{10V}{3}$

   $T_3 = V$

   ii Proof, $V = 24$
17 i   3.2 m s^{-1} ii 6
   iii 8.5   iv 1 m s^{-2}

2 Force and motion in one dimension

Prerequisite knowledge
1 $t = 0.4$ s
2 $v = 5$ m s^{-1}

Exercise 2A
1 1000 N
2 4 m s^{-2}
3 300 kg
4 35 m
5 a 0.8 m s^{-1}
   b The balls are considered as particles and so the 1 m distance does not need to include the thickness of the balls.

6	2.5 N	5		$15\,\text{m s}^{-1}$
7	2000 N	6		1.25 m
8	60 kg	7		$76.7\,\text{m s}^{-1}$
9	33 600 N	8		4.5 s
10	80 kg	9		150 m
11	15 s	10	a	5.22 s
12	80 000 kg		b	The force provided by the flare remains vertical (often the flare may be blown at an angle).
13	15 kg	11		0.085 N

Exercise 2B

1. 40 N
2. 50 N
3. 8
4. $0.625\,\text{m s}^{-2}$
5. 4950 N
6. 600 N
7. 25 s
8. 26.25 m
9. a 5340 kg
   b The air resistance is constant or the variations in air resistance are assumed to be negligible.
10. $160\,\text{m s}^{-1}$
11. 7 N
12. 136 N
13. 4000 N
14. Reduce the driving force to 125 N.

Exercise 2C

1. 700 N
2. 18 kg
3. 2 s
4. 5 m

12. $10.6\,\text{m s}^{-1}$
13. a 3.2 m
    b Proof
14. $7.56\,\text{m s}^{-1}$
15. 3.05 m
16. 109 m
17. 5.6 N

Exercise 2D

1. 1050 N
2. 354 N
3. 195 N
4. 604.8 N
5. 420 N
6. 200 N
7. 7.07 N
8. 600 N
9. a 630 N
   b The girl is being modelled as a particle, so has only one point of contact with the helicopter, otherwise there may be contact forces where her feet are on the helicopter as well as from her seat.
10. 4 N acting from the top pad pushing downwards on the parcel.

End-of-chapter review exercise 2

1. 44 m
2. 203 N
3. 40 N
4. a 26.3 m s^{-1}
   b 3.16 s
5. a 120 s
   b 50.7 m s^{-1}
6. a 302 500 N b 6480 m
   c The submarine can reach a higher speed underwater, suggesting there is not as much resistance when the submarine is underwater.
7. a 13.5 m s^{-1} b 3.93 m
8. Accelerate with force 75 N.
9. a 0.35 N b 6.24 m s^{-1}
10. 68.75 m
11. a 140 m b 84 mm
12. 3 s
13. 8.27 m
14. a Proof b 0.0609
15. i 5.66 m s^{-1} ii 0.234 s
16. i 2 s
    ii $P = 8$ m s^{-1}, $Q = 17$ m s^{-1}
17. i 13.5 N
    ii v(m s^{-1})

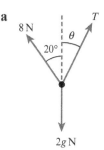

The particle enters the liquid at time 1 s with velocity −10 m s^{-1} and reaches the bottom of the container at time 1.36 s with velocity −12 m s^{-1}.

3 Forces in two dimensions

Prerequisite knowledge

1. 13 m
2. $AB = 6.13$ m, $BC = 5.14$ m
3. $\angle ABC = 33.4°$, $AB = 10.4$ m
4. $\dfrac{4}{5}$
5. $\dfrac{5}{12}$

Exercise 3A

1. i   a 11.5 N right   b 9.64 N upwards
   ii a 6.88 N left   b 9.83 N upwards
   iii a 7.52 N left   b 2.74 N downwards
   iv a 7.18 N right   b 19.7 N downwards
   v   a 7.46 N right   b 10.6 N upwards
   vi a 72.5 N right   b 33.8 N downwards
   vii a 8.47 N left   b 0.741 N downwards
   viii a 41.0 N left   b 113 N upwards

2. a $F = 10.6$ N, $F_y = 3.64$ N upwards
   b $F = 8.83$ N, $F_x = 3.73$ N right
   c $F = 12.8$ N, $\theta = 38.7°$
   d $F_y = 18.3$ N upwards, $\theta = 47.2°$
   e $F_x = 2.33$ N left, $\theta = 52.1°$

3. $F = 6.36$, $\theta = 62.4°$

4. a

   ![Diagram showing 8 N at 20° left of vertical, T at angle θ right of vertical, and 2g N downwards]

   b $T \sin\theta = 2.74$
      $T \cos\theta = 12.5$
   c $T = 12.8$, $\theta = 12.4°$

5. $F = 95.3$ N, $\theta = 70.5°$

6 a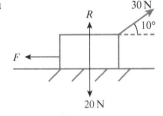

   b $F = 29.5\,\text{N}, R = 14.8\,\text{N}$

7 a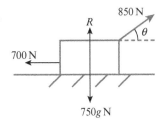

   b $\theta = 34.6°, R = 7020\,\text{N}$

8 a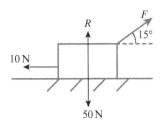

   b $F = 10.4\,\text{N}, R = 47.3\,\text{N}$

9 $T = 4260\,\text{N}, R = 51\,500\,\text{N}$

10 $F = 50\,\text{N}, G = 34\,\text{N}$

11 $T_1 = 17\,\text{N}, T_2 = 39\,\text{N}$

12 At 30° the box cannot remain on the ground. The force is not large enough to break equilibrium horizontally, so the box lifts off the ground first.

13 Proof, $F = \left(75\sqrt{3} - 100\right)$

14 $\alpha = 53.1°, \beta = 67.4°$

Exercise 3B
1 a 123 N in the given direction
  b 12.7 N in the given direction
  c 5.26 N in the opposite direction
  d 3.59 N in the opposite direction

2 a 38.1 N anticlockwise from given direction
  b 9.14 N clockwise from given direction
  c 0.212 N clockwise from given direction
  d 3.86 N anti-clockwise from given direction

3 $F = 2.90\,\text{N}, \theta = 53.5°$

4 $T = 68.4\,\text{N}, F = 53.2\,\text{N}$

5 $F = 7.76\,\text{N}, R = 29.0\,\text{N}$

6 $\theta = 17.5°, R = 38.2\,\text{N}$

7 $F = 4.57\,\text{N}, R = 18.7\,\text{N}$

8 $\theta = 42.1°, R = 80.6\,\text{N}$

9 $F = 26.6\,\text{N}, R = 76.5\,\text{N}$

10 $\theta = 33.5°, R = 155\,\text{N}$

11 $F = 20.4\,\text{N}, R = 23.0\,\text{N}$

12 a Any arrangement works. If the man holds the rod at 40°, each child can pull with force 71.0 N. If the man holds the rod at 50°, each child can pull with force 80.6 N. If the man holds the rod at 60°, each child can pull with force 89.7 N.
   b They can hold it in equilibrium provided the man holds the one at 40°.

13 a They can hold it in equilibrium if the strongest person holds the one at 10° and the next strongest holds the one at 25°.
   b They can prevent the box from moving horizontally if the strongest person holds the one at 10° and the next strongest holds the one at 25°. However, to remain in equilibrium the contact force would be negative, so the box cannot remain on the ground.

Exercise 3C
1 $\theta = 103.5°, \phi = 116.9°$

2 30 N force makes an angle of 53.1°, 40 N force makes an angle of 36.9°.

3 $T = 56.2\,\text{N}, \theta = 13.2°$

4 $\theta = 153.2°, F = 68.0\,\text{N}$

5 13.3 N at 100°, 26.2 N at 210°

6 5 N at 120°, 8.66 N at 150°

7 108.2°

8 a Up to 376 N   b Less than 220 N
  c 28.1°

Answers

9 Proof

10 Proof

Exercise 3D

1 $a = 3.88\,\text{m s}^{-2}$, $R = 45.2\,\text{N}$

2 a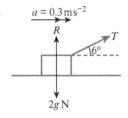

b $T = 0.603\,\text{N}$, $R = 19.9\,\text{N}$

3 $\theta = 53.1°$, $R = 84\,\text{N}$

4 $T = 58.5\,\text{N}$, $a = 0.129\,\text{m s}^{-2}$

5 $\theta = 20.2°$, $a = 3.34\,\text{m s}^{-2}$

6 a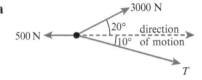

b $T = 5910\,\text{N}$, $a = 0.543\,\text{m s}^{-2}$

7 $T = 1040\,\text{N}$, and $\theta = 75°$

8 $0.433\,\text{m s}^{-2}$

9 a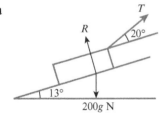

b $543\,\text{N}$

10 $F = 54.9\,\text{N}$, $a = 0.731\,\text{m s}^{-2}$

11 $\theta = 16.8°$, $a = 3.25\,\text{m s}^{-2}$

12 $5.37\,\text{s}$

13 $3.02\,\text{m s}^{-1}$

14 a $4.61\,\text{m}$

b The ball is being modelled as particle and slides up the slope.

15 $4.34\,\text{s}$

16 $1.5\,\text{s}$

17 $5.31\,\text{m}$

18 $4.14\,\text{m s}^{-1}$

19 $14.1\,\text{N}$

20 $6150\,\text{N}$

Exercise 3E

1 $a = 1.94\,\text{m s}^{-2}$ at an angle of 30.9° to the right of the positive y-direction

2 $26.6°$

3 $43.9°$ above the positive x-direction

4 $a = 1.83\,\text{m s}^{-2}$ at an angle of 14.2° above the positive x-direction

5 $15.8\,\text{N}$ at an angle of 18.4° below the positive x-direction

6 a $53.1°$

b $15.5\,\text{N}$ at an angle of 75° below the negative x-direction

7 $45°$ below the negative x-direction

8 a $a = 0.619\,\text{m s}^{-2}$ at an angle of 35.0° to the right of the direction AB

b The people continue to pull at these angles once the motion starts, otherwise the answer will only be the initial direction of motion.

9 Bearing 36.4°, $a = 0.860\,\text{m s}^{-2}$

10 The mass moves on a bearing of 088.1°, so closest to Bob.

11 Bearing 021°, $a = 0.463\,\text{m s}^{-2}$

12 $295\,\text{N}$

13 Proof

14 Akhil pulls at 10°, Ben pulls north and Khadijah pulls at 30° to give a net force of $734\,\text{N}$.

End-of-chapter review exercise 3

1 $F = 10.2\,\text{N}$, $\theta = 42.8°$

2 $17.7\,\text{N}$

3 a $170\,\text{N}$ b $0.361\,\text{m s}^{-2}$

4 $0.629\,\text{m s}^{-2}$ on a bearing of 358.8°

5 $\theta = 52.5°$, $T = 35.7\,\text{N}$

6 $0.257\,\text{m s}^{-2}$

7 a $064.6°$ b $0.629\,\text{m s}^{-2}$

8 $23.6\,\text{kg}$

9 a $2.40\,\text{s}$ b $0.784\,\text{s}$

10 a $0.576\,\text{m s}^{-1}$
   b The force is constant and the angle remains unchanged despite the motion starting.
   c $5.46\,\text{m}$

11 a $6.72\,\text{m s}^{-1}$ b $13.0\,\text{m}$

12 a $9.80\,\text{m s}^{-1}$ b $60.3\,\text{s}$

13 Proof

14 Proof

15 Resultant $= 73\,\text{N}$, direction $41.1°$ from positive x-direction

16 $AP = 6.5\,\text{N}$, $BP = 10\,\text{N}$

17 35.3

Cross-topic review exercise 1

1 $700\,\text{N}$

2 a At $t = 5\,\text{s}$, $v = 4\,\text{m s}^{-1}$ and at $t = 40\,\text{s}$, $v = 3\,\text{m s}^{-1}$
   b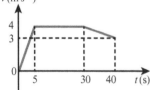

3 a $2.39\,\text{m s}^{-1}$ b $2.87\,\text{m s}^{-1}$

4 $406.4\,\text{N}$

5 a $F = 30.2\,\text{N}$, $\alpha = 34.2°$
   b $R = 36.4\,\text{N}$, $\theta = 16.4°$

6 a $528\,\text{m}$ b Proof

7 $T = 87\,\text{N}$, $F = 60\,\text{N}$

8 a $71.3°$
   b $m = 2\,\text{kg}$, angle $54.3°$

9 a $12.9\,\text{N}$ in direction AB and $7.34\,\text{N}$ perpendicular to AB above AB
   b Magnitude $14.8\,\text{N}$ at angle $29.6°$ to AB above AB

10 a $0.5\,\text{m s}^{-2}$ b $2.9°$

11 $\alpha = 84.8°$, $F = 5.52$

12 i $45\,\text{s}$
   ii $33\,\text{s}$

13 i $F = 1.90$, $R = 12.4$
   ii $0.526\,\text{kg}$

14 Tension in S_1 is $13\,\text{N}$, tension in S_2 is $20\,\text{N}$.

4 Friction

Prerequisite knowledge

1 $17\,\text{m}$

2 $12.2\,\text{m}$

3 $3.34\,\text{m s}^{-2}$

Exercise 4A

1 a $40\,\text{N}$ horizontally to the left
   b $23.5\,\text{N}$ horizontally to the right
   c $0\,\text{N}$

2 a $36.2\,\text{N}$ up the slope
   b $13.8\,\text{N}$ down the slope
   c $55.9\,\text{N}$ up the slope
   d $22.1\,\text{N}$ up the slope

3 a $204\,\text{N}$ b $193\,\text{N}$
   c $218\,\text{N}$

4 a $40\,\text{N}$ b $41.5\,\text{N}$
   c $36.8\,\text{N}$

5 $1.5\,\text{N}$

6 a $1.80\,\text{N}$ b $23.6\,\text{N}$

7 $11.4\,\text{N}$

8 $708\,\text{N}$

9 0.258

10 **a** 0.221

    b The roller is being modelled as a particle so it does not roll down the slope.

11 0.400

12 $\mu \geq 0.275$

13 1.84

14 $39.4\,\text{N} \leq T \leq 399\,\text{N}$

15 **a** 80.9° with the upward slope

    b 83° with the upward slope

16 **a** At that value of tension there will be no normal contact force, so no friction.

    b As the tension increases, the normal contact force increases, thereby increasing friction, so a smaller coefficient of friction may be enough to prevent motion.

17 $0\,\text{N} \leq T \leq 19.0\,\text{N}$ or $T \geq 198\,\text{N}$

Exercise 4B

1 **a** 140 N   **b** 14 N   **c** $0.5\,\text{m s}^{-2}$

2 **a** 29 000 N   **b** 7240 N   **c** $0.173\,\text{m s}^{-2}$

3 $0.182\,\text{m s}^{-2}$

4 **a** $1.29\,\text{m s}^{-2}$

    b The gardener will not move the wheelbarrow.

5 0.447

6 0.236

7 $3.73\,\text{m s}^{-2}$

8 0.516

9 0.268

10 $25.9\,\text{m s}^{-1}$

11 37.4 m

12 Pulling with the string gives an acceleration of $0.544\,\text{m s}^{-2}$ compared to an acceleration of $0.5\,\text{m s}^{-2}$ by pushing.

13 **a** 0.244

    b $1.19\,\text{m s}^{-2}$

14 0.448 m

15 **a** $2.97\,\text{m s}^{-2}$ **b** $0.485\,\text{m s}^{-2}$

16 **a** 13 m

    b The tension instantly falls to zero when the string breaks.

Exercise 4C

1 12.7 m, proof

2 **a** 0.954 s, proof

    b A ball would always roll, whereas a particle would not slide.

3 **a** 0.422   **b** $5.18\,\text{m s}^{-2}$

4 **a** 82.7 N   **b** $6.18\,\text{m s}^{-2}$

5 Proof, 1340 N

6 Proof, $1.60\,\text{m s}^{-2}$

7 $3.32\,\text{m s}^{-1}$

8 4.32 s

9 **a** 0.972 m

    b The ball is being modelled as a particle, so slides rather than rolls and has no thickness, so the size of the ball does not affect the height reached up the slope.

10 $9.21\,\text{m s}^{-1}$

11 18.5 m

12 0.259

13 Proof

14 0.0956

Exercise 4D

1 48 N

2 45.5 N

3 253 N

4 **a** 27.4°   **b** 0.518

5 14.3°

6 120 N

7 36.1 N

8 1610 kg

9   Proof, 61.7 N

10   31.5°

11   Proof

12   Proof

End-of-chapter review exercise 4

1   42 N

2   $22.6\,\text{N} < P < 104\,\text{N}$

3   $8.14\,\text{m s}^{-1}$

4   $1.69\,\text{m s}^{-2}$

5   $2.44\,\text{m s}^{-1}$

6   $R = 400\,\text{N}, T = 150\,\text{N}$

7   a   Proof, 3.58 N   b   Proof, $0.530\,\text{m s}^{-2}$

8   It accelerates at $0.178\,\text{m s}^{-2}$ towards the younger boy.

9   a   Proof, 61.4 N
    b   60 N at 76° to the upward slope

10   a   92.2 N   b   0.393

11   a   1.83 m   b   3.26 s

12   a $0.517\,\text{m s}^{-2}$ b $3.22\,\text{m s}^{-1}$
    c   11.7 m   d   7.27 s

13   Proof

14   a   Proof   b   Proof

15   i   Proof   ii   2.43 m

16   i   Proof   ii   Proof

17   $0.578 \leqslant P \leqslant 4.49$

5 Connected particles

Prerequisite knowledge

1   a   1.25 m   b   0.5 s

2   a   200 N
    b   Surface is horizontal, no other forces act, acceleration due to gravity is $10\,\text{m s}^{-2}$.

3   a   2 N   b $(4\sin\theta - 1)\,\text{N}$

Exercise 5A

1   60 N

2   a $0.1\,\text{m s}^{-2}$ b   130 N

3   a $0.8\,\text{m s}^{-2}$
    b   Model box as a particle so air resistance can be ignored.
    c   8 N   d   12 N

4   a   80 N   b   130 N

5   a   1700 N   b $0.5\,\text{m s}^{-2}$

6   a   i   Upper rod 240 N, lower rod 120 N
      ii   Upper rod 240 N, lower rod 160 N
    b   The masses of the rods are negligible, the second rod is vertical, the buckets of water can be modelled as particles.

7   From top: 250 N, 240 N, 230 N, 220 N, 210 N, 200 N, 190 N, 180 N, 170 N, 160 N, 150 N

8   25 N, 25 N, 40 N

9   70 000 N

10   344 N tension

11   4 N thrust

12   Proof

Exercise 5B

1   a   Tension 30 N, friction 30 N
    b   Tensions 50 N, 30 N, friction 20 N
    c   Tension 30 N, friction 10 N
    d   Tension 20 N, friction with horizontal surface 20 N

2   a   3 s
    b   The rope is modelled as a light inextensible string and the buckets as particles.

3   30 N, 4 N, 6 N

4   a   0.253 s   b $1.58\,\text{m s}^{-1}$
    c   1.08 s

5   a   1.2 s   b   1.4 s

6   a   10.6 N   b   0.44 m

7 4.2 N, 1.2 N
8 6
9 1
10 a $1\,\text{m s}^{-2}$ b 36 N, 33 N
11 a $\dfrac{3(1+\sqrt{3})}{4}\,\text{N}$ b 0.475 N, 0.275 N
12 a There are two lengths of string at the cylinder, so the distance moved by the cylinder in a given time is half the distance moved by the box, hence the speed and the magnitude of the acceleration are also half those of the box.
   b $2.5\,\text{m s}^{-2}$ downwards
13 a Strings are light, inextensible and hang vertically.
   b 5.5 N in each   c 2.5 N in each
   d Upper unchanged, lower changed to 3 N.

Exercise 5C
1 a 206 N   b 465 kg
2 a 3300 N   b 3200 N
   c 3100 N
3 6
4 a 5200 N   b 416 N
5 a 405 kg   b 368 N
   c 158 N   d 52.5 N
6 a $2.5\,\text{m s}^{-2}$ b 417 N
7 a $(M+m)(10+a)\,\text{N}$ b $m(10+a)\,\text{N}$
8 a 32.1 N   b Proof
9 a i 616 kg   ii 797 kg
   b 8
10 10 000 N
11 a $2\,\text{m s}^{-2}$, upwards for A and downwards for B
   b $0.4\,\text{m s}^{-2}$ upwards
   c A $2.4\,\text{m s}^{-2}$ upwards, B $1.6\,\text{m s}^{-2}$ downwards

End-of-chapter review exercise 5
1 a 1 s   b 84 N
2 a 100 N tension b 5 N tension
3 a $9.5\,\text{m s}^{-2}$ b 7.5 N
4 a $1\,\text{m s}^{-1}$ b 0.9 s
5 a $2.5\,\text{m s}^{-2}$ b 2.7 m
6 2.5 s
7 a Pulleys are smooth, crate and ball are modelled as particles, rope is light and inextensible. If the pulleys are not smooth they might 'stick' and the tension in the rope might be different on the two sides of the pulleys, but the second pulley would (probably) still not move.
   b $24\sqrt{2} = 33.9$
8 i $T_A = 2.5 + 0.25a$, $T_B = 7.5 - 0.75a$
   ii Proof
   iii $1.2\,\text{m s}^{-1}$ iv $6\,\text{m s}^{-2}$
9 Proof
10 i $6\,\text{m s}^{-2}$ ii $3.29\,\text{m s}^{-1}$
11 i $2.93\,\text{m s}^{-1}$ ii 0.84 m
12 a $1.2\,\text{m s}^{-1}$ b   Proof
   c   0.0868   d $2.01\,\text{m s}^{-1}$

6 General motion in a straight line

Prerequisite knowledge
1 a $14\,\text{m s}^{-1}$ b 33 m
   c 5 s
2 a $15x^2 - 60$, $(-2, 82)$, $(2, -78)$
   b $\dfrac{5}{4}x^4 - 30x^2 + 2x + c$, -96

Exercise 6A
1 $20\,\text{m s}^{-1}$
2 $86\,\text{m s}^{-1}$
3 a Ball is modelled as a particle, air resistance can be ignored.
   b $20\,\text{m s}^{-1}$ c $0\,\text{m s}^{-1}$
4 a $0\,\text{m s}^{-1}$ b $0\,\text{m s}^{-1}$
   c 12 units per second
5 a Proof
   b 7 m

6 a 1.6 s b 3.2 m
7 a Proof
   b 12 m
8 a 320 m b 36 m s^{-1}
   c 39.2 m s^{-1} d 12.8 m
9 2.7 m
10 a s is continuous at $t = 4$,
      so $5 \times 4^2 = A\sqrt{4} + B \times 4$
      $80 = 2A + 4B$
      $40 = A + 2B$
   b Proof
   c $A + 5B = 5C + 6$, $A + 10B = 10C$
   d 23
11 a Proof
   b 4.04 m s^{-1}
12 a Proof b 6.74 m

Exercise 6B

1 $t = 2$
2 16 m s^{-2}
3 a -10 m s^{-2}
   b This is the acceleration due to gravity. It is negative because the upward direction is positive in this question.
4 a 5 m s^{-2} b 6 m s^{-2}
   c 7 m s^{-2}
5 a 11 b Proof
   c $-\dfrac{11}{75}$ m s^{-2} d Proof
6 a $v = 6 - 2t$ m s^{-1}
   b It starts with speed 6 m s^{-1} but slows down and then stops and returns back the way it came, speeding up all the time.
   c $t = 3$ s
   d $a = -2$ m s^{-2} (constant acceleration)
7 250 m s^{-1}

8 $A = -\dfrac{2}{9}$, $B = \dfrac{2}{3}$, $C = 2$
9 a 21.3 m s^{-1} b 5.33 s
10 a 5 m s^{-1} b 9 m s^{-1}
   c -1 m s^{-2}
11 a 10 minutes
   b 75.2 km h^{-1} (or 20.9 m s^{-1})
   c 12.5 km

Exercise 6C

1 $-\dfrac{80}{3}$
2 14.25 m
3 20 m
4 a 0 m b 1.33 m
   c $\dfrac{20}{3}$ m
5 a 29.9 m
   b Smaller
6 a 1 s b 7 cm
   c 1.18 s
   d There is no resistance on the downward journey, so, for example, the ball bearing does not touch the sides of the hole it has made. Improve model by incorporating a factor to represent friction.
7 a -43.8 m b 84.3 m
8 a 87.5 m b 271 m
9 a Proof
   b 294 m
10 a 16 b 57.3
11 a First car 0 m s^{-1}, second car 16 m s^{-1}
   b 281 m
   c 42.7 m d 12.3
12 a $k = 1$ b Proof
   c 6.55 m s^{-1} d $T = 22$ s
   e 95.3 m

Exercise 6D
1. $3.2\,\text{m s}^{-1}$
2. $14.2\,\text{m s}^{-1}$
3. a $18.5\,\text{m s}^{-1}$ b $-1.2\,\text{m}$
   c Towards
4. a Proof b $5.78\,\text{m}$
   c Proof d $3.43\,\text{m s}^{-1}$
5. $4.98\,\text{m}$
6. $2.52\,\text{m}$
7. a Proof b $2\,\text{m}$
8. 6
9. $1690\,\text{m}$
10. a $19.8\,\text{m s}^{-1}$ b $7.6\,\text{s}$
    c $42.7\,\text{m}$ d Proof
11. a $80.0\,\text{m}$ b $39.9\,\text{m}$
    c $2.66\,\text{s}$
12. a Proof b Proof
    c According to the model there is still air resistance when the ball comes to the end of the alley, but the ball will stop when it reaches the end of the skittle alley.

End-of-chapter review exercise 6
1. $26.7\,\text{m}$
2. a $0.125\,\text{m s}^{-2}$ b $111\,\text{m}$
3. a $12\,\text{m s}^{-1},\ 3\,\text{m s}^{-1}$ b $175\,\text{m}$
4. a $8\,\text{m s}^{-1}$ b Proof
5. $23.4\,\text{km}$
6. a $4.5\,\text{m s}^{-1}$ b $40\,\text{s}$
7. a Proof b Proof
   c $69.2\,\text{m}$
   d The gradient of the acceleration–time graph changes suddenly at $t = 1$.
8. i $\frac{1}{6}\,\text{m s}^{-2}$ ii $71.3\,\text{m}$
9. a $6.5\,\text{m s}^{-1}$ b $63.8\,\text{m}$
   c $3.06\,\text{s}$
10. i $6\,\text{m s}^{-1},\ 0\,\text{m s}^{-1}$
    ii $583\,\text{m}$
11. i $A = 4$, proof
    ii $\left(450 - \frac{3375}{t}\right)\text{m}$
    iii $5.4\,\text{m s}^{-1}$
12. a Proof b $3\,\text{m s}^{-1}$
    c 2.1 d Proof
    e Proof

Cross-topic review exercise 2
1. a Tension = $48\,\text{N}$, acceleration = $2\,\text{m s}^{-2}$
   b $96\,\text{N}$
2. On the point of slipping down, $\mu = 0.336$
3. $0.580\,\text{s}$
4. Braking force = $6\,\text{N}$, compression of $2\,\text{N}$ in the tow-bar.
5. a $7.5\,\text{m s}^{-1}$ b $59\,\text{m}$
6. a $1.59\,\text{m}$ b $2.94\,\text{s}$
7. a $24.8\,\text{m}$ b $3\,\text{s}$
8. a $2.68\,\text{m}$ b $1.99\,\text{s}$
9. a $2\,\text{s}$ b $9.12\,\text{m}$
   c $14\,\text{m}$
10. $a = \frac{g}{3} = 3.33\,\text{m s}^{-2},\ T = \frac{20g}{3} = 66.7\,\text{N}$
11. i $a = 0.4\,\text{m s}^{-2}$, proof ii $\mu = 0.321$
12. i $t < 2.5$ ii $27\,\text{m}$
    iii $t = 2.31$ and $t = 5.19$
13. $0.8\,\text{m}$

7 Momentum
Prerequisite knowledge
1. a Proof b $7.5\,\text{m s}^{-1}$
   c $10.6\,\text{m s}^{-1}$
2. a $7.5\,\text{m s}^{-1}$ b $12.4\,\text{m s}^{-1}$

Exercise 7A
1. 80 N s
2. 33 000 N s
3. 2.85 N s
4. 0.056 N s
5. 36 N s
6. a $7\,\text{m s}^{-1}$ b 245 N s
7. a $6\,\text{m s}^{-1}$ b 12 N s
8. a 0.45 m
   b Ball is modelled as a particle with no size and it is assumed there is no air resistance. Air resistance would slow the ball, so it would have a smaller velocity when it reaches the ground and consequently a smaller velocity after the bounce. If the ball has size then the centre of the ball does not reach the ground so the distance travelled is reduced and again this will reduce the velocity after the bounce. Reduced rebound velocity will reduce the height reached.
9. 0.125 N s
10. Proof
11. $5\,\text{m s}^{-1}$
12. 1.33 m

Exercise 7B
1. 40 kg
2. $2\,\text{m s}^{-1}$
3. 50 kg
4. 9 : 11
5. $0.5\,\text{m s}^{-1}$
6. $2\,\text{m s}^{-1}$ in the same direction as C (the original direction of travel for A)
7. a $6\,\text{m s}^{-1}$ in the opposite direction to Jayne
   b Motion is in a straight line. Jayne just stands and does not 'push off' with her skates. Jayne and chair can be modelled as particles.
8. 4.9 kg
9. a $2.5\,\text{m s}^{-1}$ b It is horizontal.
10. a Proof   b Proof
    c $1880\,\text{m s}^{-1}$
11. a Proof   b $2(\alpha - 1)$
12. a $0.75\,\text{m s}^{-1}$, $0.5\,\text{m s}^{-1}$, $3\,\text{m s}^{-1}$, $2\,\text{m s}^{-1}$
    b Opposite.

End-of-chapter review exercise 7
1. 2
2. $0.8\,\text{m s}^{-1}$
3. 90.9 s
4. 0.6
5. $\frac{1}{3}\,\text{m s}^{-1}$
6. a $(5m + 0.4)\,\text{N s}$ b 0.1
7. $1.4\,\text{m s}^{-1}$
8. a X: 0.45 N s, Y: −0.05 N s   b Proof
9. $1.95 < m < 2.5$
10. a 6.84 N s
    b Height reached after second bounce 1.18 m, after third bounce 0.957 m.
    c Ball can be modelled as a particle, so ball has no size, and there is no air resistance.
11. a 0.0225 N s
    b Let the original direction of travel for X be the positive direction.
       Total momentum before impact = − 0.0072 N s
       This is negative so at least one ball must be travelling in the negative direction after impact.
       Y cannot pass through X, so X must reverse its direction of travel.
    c $-0.045\,\text{m s}^{-1}$
12. Proof

8 Work and energy

Prerequisite knowledge
1. **a** 20 N **b** 34.6 N
2. 6 N up the slope
3. **a** 18 N down the slope **b** 4.5 m s^{-2}
4. 56.3 cm

Exercise 8A
1. 60 J
2. **a** 100 J **b** 76.6 J
3. **a** −0.8 J **b** 0.8 J
   c 0 J
4. 2400 J
5. **a** Friction from the edge of the canal, resistance from the water and some air resistance
   b 2000 J
6. **a i** 5910 J **ii** 5640 J
   b Proof
   c The component of the tension perpendicular to the direction of motion is more than double in the second situation (53.8 N) compared with the first (26.0 N) so the frictional resistance will be greater.
7. **a** 6 J **b** 17.3 J
   c 0 J **d** 0 J
   e 11.3 J
8. **a** 20 J **b** 259 J
   c 0 J **d** 239 J
9. **a** 400 J **b** 20 J
   c 259 J **d** 0 J
   e 121 J
10. **a** 5 J **b** 3 J
    c 0 J **d** 2 J
11. 283 J
12. 1.73 J

Exercise 8B
1. 320 J
2. 363 000 J or 363 kJ
3. 71.3 J
4. 54 J
5. **a** 14.7 m s^{-1} **b** 216 J
6. **a** 13 m s^{-1} **b** 80 J
7. 7 m s^{-1}
8. 2.5 m
9. **a** 1250 J
   b No difference, provided the speed is still constant.
10. **a** 5.63×10^{15} J
    b Fuel will be used while the rocket is accelerating, so the mass will decrease.
11. **a** Proof **b** $\dfrac{13}{6}$ or 2.17 m s^{-1}
12. Momentum is conserved
    $mu_A - mu_B = -mv_A + mv_B$
    so $u_A + v_A = u_B + v_B$
    Kinetic energy is conserved
    $0.5mu_A^2 + 0.5mu_B^2 = 0.5mv_A^2 + 0.5mv_B^2$
    so $u_A^2 - v_A^2 = v_B^2 - u_B^2$
    $(u_A - v_A)(u_A + v_A) = (v_B - u_B)(v_B + u_B)$
    But $u_A + v_A = v_B + u_B$
    so $u_A - v_A = v_B - u_B$
    Add: $(u_A + v_A) + (u_A - v_A) = (v_B + u_B) + (v_B - u_B)$
    so $2u_A = 2v_B$
    $u_A = v_B$ and $v_A = u_B$
13. **a** 0.12 J **b** 0 J **c** 0.21 J

Exercise 8C
1. 100 J
2. 600 J decrease
3. 0.399 J
4. **a** 500 J **b** 500 J **c** 500 J

5   20.6 J
6   5670 J
7   40 kg
8   a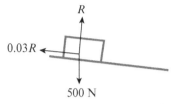
    b   2.10 m   c   131 J
9   a   Proof   b $\dfrac{v^2}{11.32}$ m   c $37.5v^2$ J
    d   The boy is modelled as a particle, this means air resistance is ignored. Air resistance would slow the boy down, so the slope would be longer and the loss in GPE would be greater than the values given.

    The slope is modelled as a straight line. In reality it would flatten out towards the bottom, so the boy would slow down while travelling horizontally, his speed at the bottom of the descent would be greater than v and the loss in GPE would be greater than the value given.
10 −60 J
11 0.075 J
12 Proof

End-of-chapter review exercise 8
1   24.3
2   a   0.24 J   b   0.24 J
3   a $0.978\,\text{m s}^{-2}$ b   321 kJ
4   a   300 N   b   Proof
    c   The push force and resistance are constant.
5   i   750 J   ii Proof
6   a   6.25 J decrease   b   6.25 J increase
    c   No difference to the numerical answers, but it would affect how far the ball travels horizontally, the height the ball reaches and also the angle that the path of the ball makes with the vertical when the ball passes through the hoop.

7   i   1.7   ii $30\,\text{m s}^{-1}$
8   a $2\,\text{m s}^{-2}$ b   2.25 m
    c   90 J
9   a   20 000 J   b   10 000 J
    c   50 m   d   10 000 J
10 a   640 N   b   4 m
    c   30°   d $6.12\,\text{m s}^{-1}$
11 a $1.50\,\text{m s}^{-2}$ parallel to the slope and down the slope.
    b   8.35 m   c   334 J
    d   Proof
12 a   The only force acting on Jack during the flight is his weight, which is vertical, so there is no horizontal resultant force and hence no horizontal acceleration.
    b $13.2\,\text{m s}^{-1}$
    c   1060 J
    d   1.51 m
    e   He could easily slide off the trampoline.
    f   He would bounce up to quite a height and could bounce several times before coming to rest.

9 The work–energy principle and power

Prerequisite knowledge
1   12.5 J
2   a   150 J   b   15 J   c   125 J

Exercise 9A
1   a   60 J   b   110 J
    c   50 J   d $2\,\text{m s}^{-1}$
2   $1.39\,\text{m s}^{-1}$
3   a   400 J   b   400 J   c   0.80 m
4   a   2550 J   b $7.75\,\text{m s}^{-1}$ c $7.75\,\text{m s}^{-1}$
5   a $10.2\,\text{m s}^{-1}$ b   25 m
6   63 N
7   5 m

Exercise 9A

1. **a** 60 J **b** 110 J
   c 50 J **d** 2 m s^{-1}
2. 1.39 m s^{-1}
3. **a** 400 J **b** 400 J
   c 0.80 m
4. **a** 2550 J **b** 7.75 m s^{-1}
   c 7.75 m s^{-1}
5. **a** 10.2 m s^{-1} **b** 25 m
6. 63 N
7. 5 m
8. **a** 0.498 N
   b Very small resistance force so grass is very slippery, perhaps the grass is wet.
9. **a** 800 J **b** 7.14 m
   c 3.14 m
10. If the initial speed is u m s^{-1} and the final speed is v m s^{-1} then $v^2 - u^2 = 300$. According to the driver, $v < 30$ so u must be less than 24.5.
11. **a** Proof
    b Proof
12. **a** 15 J **b** Proof
    c 5.83 m s^{-1} **d** 5.48 m s^{-1}
    e 67.2° **f** 6.32 m s^{-1}

Exercise 9B

1. 6.33 m s^{-1}
2. 5.35 m s^{-1}
3. **a** 0.57 J
   b 49.8 m s^{-1} = 179 km h^{-1}
4. 2 m s^{-1}
5. 5 m s^{-1}
6. 2.04 m s^{-1}
7. $0.05 v^2$
8. **a** $\sqrt{(200 + u^2)}$ m s^{-1}
   b Diver modelled as a particle so no air resistance, no spin etc. End of the board assumed to be 10 m above the water at take off, but if it is a flexible board it may be less (or more) than 10 m.
9. **a** 14.3 m s^{-1}
   b The only force acting is the weight, which is vertically downwards, so there is no horizontal component to the acceleration.
   c Proof
10. **a** **i** $10mh$ J **ii** $0.5mv^2$ J
    iii $(0.5Mv^2 + 10Mh \sin\theta)$ J
    b Proof
11. **a** **i** 7 m s^{-1} **ii** 11 m s^{-1}
    b The surface is smooth.
    c No difference.
12. **a** $\sqrt{20} = 4.47$ m s^{-1} **b** 75.5°

Exercise 9C

1. **a** 24 m **b** 1200 J
   c Heat
2. 0.30
3. 24 J
4. 1200 N
5. 1.71 N
6. 20 m
7. $\left(10 + \frac{1}{21} u^2\right)$ m
8. **a** 22.5 m s^{-1} **b** 1.44 m
9. **a** 0.185 N **b** 35 m s^{-1} = 126 km h^{-1}
   c Proof
10. **a** 4.79 **b** 5.32 cm
11. **a** 2 N **b** $\dfrac{61 - v^2}{6}$ N
12. **a** 2.1 J **b** Proof

10 a $50\,\mathrm{m\,s^{-1}}$ b Resistance is now 30 N.
11 2450
12 a Energy dissipated = $625(300 - v^2)\,\mathrm{J}$,
       distance travelled = $1.25(300 - v^2)\,\mathrm{m}$
    b $\left(\dfrac{20}{v} - 0.4\right)\mathrm{m\,s^{-2}}$, Proof
    c Proof, $14.8\,\mathrm{m\,s^{-1}}$

End-of-chapter review exercise 9
1 $P = 28\,500$, $R = 750$
2 a $\dfrac{40m}{m+2}\,\mathrm{N}$ b $\dfrac{0.6(m+10)}{(m+2)}\,\mathrm{m}$
3 $v = \sqrt{\dfrac{x}{5}}$
4 a $26.3\,\mathrm{m\,s^{-1}}$
   b The speed would decrease
   c The speed would decrease
5 2.50
6 a $20\,\mathrm{m\,s^{-1}}$ b $39\,\mathrm{kW}$
7 $3.03\,\mathrm{m\,s^{-1}}$
8 a 1200 N b 0.025
9 i Gain in KE = 4030 kJ,
       loss in PE = $42\,000\sin\theta$ kJ
   ii 3.26°
10 a $0.625(500 - T)\,\mathrm{J}$ b $(0.0991T - 6.25)\,\mathrm{m\,s^{-1}}$
11 i $10.5v^2\,\mathrm{J}$
   ii a $135x\,\mathrm{J}$ b $25x\,\mathrm{J}$
   iii Proof
12 a $12.5x\,\mathrm{J}$ b $10\sqrt{2}x\,\mathrm{J} = 14.1x\,\mathrm{J}$
   c $0.906\,\mathrm{m\,s^{-1}}$
   d Work done by tension in pulling W up slope AB is cancelled by work done against tension when X slides down BC.

Cross-topic review exercise 3
1 2.5 N s upwards
2 58.4 J

3 Yes, there will be a third collision, because object B is moving faster than object A and will catch it up.
4 a 50 b $6\,\mathrm{m\,s^{-1}}$
5 a $1.5\,\mathrm{m\,s^{-1}}$
   b $3\,\mathrm{m\,s^{-1}}$ in the opposite direction from the coalesced particle
6 a 300 m b 991 m
7 a $8\,\mathrm{m\,s^{-1}}$ b 48 J
   c 0.8 m
8 a 750 b 250 m
9 a Proof, $4\,\mathrm{m\,s^{-1}}$ b 1.5 m
10 i 514 kW ii $31.7\,\mathrm{m\,s^{-1}}$
11 i 1440 J ii 3080 J
   iii 4520 J
12 i 12 kW ii $0.132\,\mathrm{m\,s^{-2}}$
13 i 985 J ii $5.55\,\mathrm{m\,s^{-1}}$
   iii 273 W iv 54.4 s

Practice exam-style paper
1 Proof, $F = 6.61\,\mathrm{N}$
2 a 1875 N b 375 kJ
3 a 0.72 kg b 10.8 J
4 a 225 m b 25 s
5 a $V = 4\,\mathrm{m\,s^{-1}}$ b Proof
   c 28.5 s
6 a 0.75 N b 1.27 m
7 a $0.775\,\mathrm{m\,s^{-2}}$ b 1.16 J
   c 19.7 J d 36.0 N

Glossary

A

Acceleration: rate of change of velocity

Angle of friction: the angle between the normal contact force and the total contact force when friction is limiting

C

Coalesce: when two bodies join together in an impact and continue as one object; the opposite of an explosion

Coefficient of friction: the ratio between the frictional force and the normal contact force when friction is limiting

Components: the parts of a force acting parallel to given axes, usually two perpendicular axes

Compression: force in a rod or other connecting object, but not a string, which provides a force in the direction of the rod towards the object it is connected to

Connected objects: objects that are attached together with forces acting between them

Conservative force: a force for which the work done in moving an object between two points is independent of the path taken

Conserved: unchanged, as in 'total momentum is conserved in an impact'

Contact force: the combined effect of two objects touching, comprising two components: the normal contact force and friction

D

Displacement: distance relative to a fixed point or origin in a given direction

Dissipated: mechanical energy lost by being converted into non-mechanical energy, such as heat, sound and light

Distance: length of path between two points

E

Equilibrium: state of an object when there is no net force acting on it

Explosion: when a single object splits into two or more separate parts; the opposite of coalescence

F

Force: influence on an object that can alter its motion

Friction: force between two surfaces acting parallel to the contact between the surfaces, as a result of the roughness of the surfaces in contact

G

Gravitational potential energy (GPE) (or potential energy (PE)): the energy that a body possesses because of its position (in a gravitational field). The potential energy is the product of the weight and the height. It is a scalar quantity measured in joules (J)

Gravity: attraction between two objects as a result of their masses, usually thought of as a force acting on an object towards the Earth

I

Impact: a collision or other interaction between two bodies

Instantaneous acceleration: the acceleration at an instant, which is the gradient at a point on a velocity–time graph; usually just referred to as acceleration

Instantaneous velocity: the velocity at an instant, which is the gradient at a point on a displacement–time graph; usually just referred to as velocity

K

Kinetic energy (or linear kinetic energy): the energy that a body possesses because of its motion; calculated as half the product of the mass and the square of the speed; a scalar quantity, measured in joules (J)

L

Light: having no, or negligible, mass

Limiting equilibrium: when friction is at its maximum possible value but there is no net force on the object

Line of action: the direction in which a force acts

Line of greatest slope: the steepest path up or down a surface which is at an angle to the horizontal

M

Momentum, or linear momentum: the product of the mass and the velocity of an object; a vector quantity, measured in N s; its direction is the same as the direction of the velocity.

N

Negligible: small enough to be ignored for the purposes of the mathematical model

Newton's first law: the principle that an object continues moving in the same direction at the same speed unless a net force acts on the object

Newton's second law: the principle that the rate of change of momentum is proportional to force acting on an object, which leads to the equation $F = ma$ in the case where mass is constant

Newton's third law: the principle that for every action there is an equal and opposite reaction

Non-conservative force: any force for which the work done in moving a particle between two points is different for different paths taken

Normal contact force: influence of one object on another through being in contact; the force acts perpendicular to the touching surfaces

O

'On the point of slipping': state of an object when friction is limiting so that any increase in the force applied to the object will cause it to move

Origin: reference point from which displacement is measured

P

Power: the rate of doing work, measured in watts; the power generated by the engine of a vehicle is the product of the driving force and the speed at which the vehicle is moving

R

Reaction force: alternative name for normal contact force

Resistance: force opposing motion possibly caused by the air or other medium through which the object moves

Resolving: process of splitting forces into components in given (usually perpendicular) directions

Resultant: single force equivalent to the net total of other forces

Rod: any light rigid connector joining two objects; it can be in tension or in thrust

Rough: having friction

S

Scalar: quantity having a numerical value but no assigned direction

Smooth: having no friction

Smooth pulley: a pulley for which the magnitude of the tension in a string passed over it is the same on each side of the pulley

Speed: rate of moving over a distance

String: any flexible connector joining two objects; it can be in tension but not in thrust; it is assumed to be light and inextensible (does not stretch)

T

Tension: force in a string, or other connecting object, which provides a force in the direction of the string away from the object it is connected to

Thrust: the force provided by, for example, a rod when under compression, acting along the rod towards an object

V

Vector: quantity having a numerical value in an assigned direction, which may be negative

Velocity: rate of change of displacement

W

Work: the work done by a force that causes an object to move along the line of action of the force is the product of the magnitude of the force and the distance the object moves in the direction of the force; a scalar quantity, measured in joules (J)

Work done against gravity: the work done by the weight of a body when the body is raised vertically; if the body has mass m and rises through a vertical height h then the work done against gravity is mgh; equal to the *increase* in potential energy

Work done by gravity: the work done by the weight of a body when the body falls vertically; if the body has mass m and falls through a vertical height h then the work done by gravity is mgh; equal to the *decrease* in potential energy

Work–energy principle: for any motion the increase in kinetic energy is equal to the work done by all forces or the increase in mechanical energy is equal to the work done by all forces (excluding weight)

Index

acceleration 9–10, 37
   as the derivative of velocity with respect to time 136–9
   due to gravity (g) 42–5
   equations of constant acceleration 11–13
   finding the direction of 72–5
   and friction 91
   instantaneous 136
   modelling assumptions 23
   Newton's second law of motion 37–8, 39, 67–9
   from velocity–time graphs 22–3
   see also non-constant acceleration
air resistance 36, 42
angle of friction 101–4
Archimedes 117
area under a velocity–time graph 23
Aristotle 36, 37
average speed 3, 5
average velocity 4–5

butterfly effect 48

catenary curve 55
changes of direction 134
chaos theory 48
coalescence 165
coefficient of friction (μ) 85
   experimental determination of 102
   relationship to angle of friction 101
collisions 163–4
   coalescence and explosions 165
   modelling assumptions 28, 164, 167
communication vi
components of a vector 54–6, 59–61
compression 36
connected objects 110
   modelling assumptions 120
   in moving lifts 122–4
   by rods 111–14
   by strings 116–20
conservation of mechanical energy
   with conservative forces 196–8
   with non-conservative forces 199–202
conservation of momentum 163–7
conservative forces 196
contacts, rough and smooth 85

deceleration 9, 137
   see also acceleration
Descartes, René 161
direction, changes of 134
displacement 3, 4–5
   difference from distance travelled 143
   as the integral of velocity with respect to time 141–7
   relationship to acceleration 11
   from velocity–time graphs 22
displacement–time graphs 15–19, 132
   instantaneous changes in gradient 27
dissipation of energy 199
distance 2, 4–5
   difference from displacement 143
driving force 205–7
dynamics 2

energy 173–4
   conservation of 196–202
   gravitational potential 183
   kinetic 183
   mechanical and non-mechanical 191
   modelling assumptions 183
   work–energy principle 189–94
energy conversions
   with conservative forces 196–8
   with non-conservative forces 199–202
energy dissipation 199, 205
equations of constant acceleration 11–13
equilibrium 39, 55–6, 86
   limiting 85
   triangle of forces and Lami's theorem 63–6
Euler, Leonhard 93
explosions 165

force diagrams 39–41
forces 36
   combinations of 39–41
   conservative 196
   driving force 205–7
   gravity 42–5
   Lami's theorem 63
   line of action 174, 175–6
   modelling assumptions 37, 68

Newton's laws of motion 36–8, 39, 110–11
   non-conservative 199
   non-equilibrium problems 67–9, 72–5
   normal contact force 47–9
   resolution of 54–6, 59–61
   resultants 72–5
   triangle of 63–6
   work done against 177–8
   work done by 174–6
   see also friction
friction 36, 84–8
   angle of 101–4
   changes of direction 96–9
   coefficient of 85, 101, 102
   direction of action 85–6
   and fluid dynamics 92
   limit of 91–4
   limiting value 85, 86, 91
   modelling assumptions 92, 97
   moving objects 91
   as part of the contact force 85

Galileo Galilei 42
gradients
   displacement–time graphs 16
   instantaneous changes 28
   velocity–time graphs 22
graphs
   with discontinuities 27–9
   displacement–time 15–19, 132
   velocity–time 22–5, 142
gravitational potential energy (GPE) 183
   work–energy principle 191–4
gravity
   acceleration due to (g) 36, 42–5
   work done by or against 175, 177–8

Halley, Edmund 134
height, changes in
   work done 183
   work–energy principle 191–4

impact 163–4
   coalescence and explosions 165–7
   modelling assumptions 28, 164, 167
instantaneous acceleration 136
instantaneous velocity 132

Joule, James Prescott 174
joules (J) 174

kinetic energy 180
   modelling assumptions 183
   work–energy principle 189–94

Lami's theorem 63–6
Leibnitz, Gottfried 199
lifts, objects in 122–4
limiting equilibrium 85
line of action of a force 59, 174
   different direction from line of motion 175–6
line of greatest slope 86

mechanical energy 189, 191
   conservation of 196–202
   work–energy principle 191–4
Mechanics 2
modelling vi–vii
modelling assumptions 6–7
   connected objects 120
   forces 37
   friction 92, 97
   g, value of 43
   impact 164, 167
   instantaneous changes in velocity 28
   kinetic energy 183
   with non-constant acceleration 150
   non-equilibrium problems 68
   normal contact force 48
   power 205
   velocity–time graphs 23
momentum 160–2
   conservation of 163–7
motion, Newton's laws of 36–8, 39, 110–11
multi-stage problems
   and displacement–time graphs 15–19
   and velocity–time graphs 22–5

Navier-Stokes equations 93
Newton, Isaac 36, 134
Newton's cradle 163
Newton's first law 37

Newton's second law 37–8, 39, 67–9
   application to connected objects 111–14, 116, 118, 123–4
Newton's third law 110–11
non-conservative forces 199
non-constant acceleration 131
   displacement 141–7
   modelling assumptions 150
   velocity 132–4, 150–2
non-gravitational resistance 178
normal contact force (reaction force) 47–9
   and friction 85
   modelling assumptions 48
   Newton's third law 110–11

'on the point of slipping' 85
origin 3

Poincaré, Henri 48
potential energy (gravitational potential energy) 183
power 189, 204–7
   modelling assumptions 205
problem solving vi
pulleys 116
   modelling assumptions 120

reaction force (normal cvontact force) 47–9
   and friction 85
   modelling assumptions 48
   Newton's third law 110–11
resistance forces 37
   non-gravitational 178
   see also air resistance; friction
resolving forces 54–6
   in directions other than horizontal and vertical 59–61
resultant forces 39–41, 72–5
rods
   modelling assumptions 120
   objects connected by 111–14
rough contacts 85

scalar quantities 2
smooth contacts 85

smooth pulleys 116
   modelling assumptions 120
speed 3, 4–5
statics 2
stationary points 134
strings
   modelling assumptions 120
   objects connected by 116–20
suvat equations 11
Système Internationale (SI) system of units 161

tension 36
   in rods 111–14
   in strings 116–20
three-body problem 48
thrust 36, 111–13
total contact force 85, 101
triangle of forces 63–6

undergroundmathematics website vii

vectors 3
   components of 54–6
velocity 3–5
   as the derivative of displacement with respect to time 132–4
   from displacement–time graphs 16
   instantaneous 132
   instantaneous changes 27–9
   as the integral of acceleration with respect to time 150–2
   modelling assumptions 28
   relationship to acceleration 9, 11
   sign of 137
velocity–time graphs 22–5, 142
   discontinuities 27
   modelling assumptions 23

watts (W) 204
weight 42–3, 196
work 173
work done against a force 177–8
work done by a force 174–6
work done by or against gravity 175, 177–8
work–energy principle 189–4